International Conference Automatics, Robotics and Artificial Intelligence 2024

IOP Conference Series: Materials Science and Engineering
Volume 1317

Sozopol, Bulgaria
14-16 June 2024

ISBN: 979-8-3313-0781-3
ISSN: 1757-8981

Printed from e-media with permission by:

Curran Associates, Inc.
57 Morehouse Lane
Red Hook, NY 12571

Institute of Physics
Dirac House, Temple Back
Bristol BS1 6BE UK

Phone: 44 1 17 929 7481
Fax: 44 1 17 920 0979

techtracking@iop.org

Additional copies of this publication are available from:

Curran Associates, Inc.
57 Morehouse Lane
Red Hook, NY 12571 USA
Phone: 845-758-0400
Fax: 845-758-2633
Email: curran@proceedings.com
Web: www.proceedings.com

TABLE OF CONTENTS

IOP Conf. Series: Materials Science and Engineering **1317** (2024) 011001 doi:10.1088/1757-899X/1317/1/011001

Preface

The 2nd International Conference Automatics, Robotics and Artificial Intelligence ICARAI 2024 was successfully held at Sozopol, Bulgaria from June 14th to 16th, 2024. The ICARAI continues the 12 - year tradition of the International Conference Automatics.

The ICARAI' 2024 provides an opportunity to discuss current problems, exchange ideas and innovative solutions in various fields, such as automatic control, intelligent systems, cyber-physical systems in automation, neural networks, fuzzy logic, information and measurement systems, bio-informatics, biosensor systems and others. In parallel with the sessions, discussions on problems related to the connections between education, science and industry are also organized. In this way, the participants discuss the possibilities for improving the interaction and cooperation between scientific circles and business.

The program of the conference is available at:

https://e-university.tu-sofia.bg/e-conf/index.php?r=754

The papers presented at the conference have passed a double peer review process and we are proud to provide the collection to readers of the Journal Materials Science and Engineering – Conference Series. All papers are written at a high scientific level and represent the result of the collective efforts and work of the authors, who are constantly expanding the boundaries of science and technology.

The conference Organizing Committee of ICARAI' 2024 express its sincere gratitude to all those who supported the conference - to the authors who choose the forum for the publication of their scientific research, to the participants who attended the conference in person and/or online, to the reviewers for their honest opinions and constructive critical comments, contributed to raising the level of the scientific articles.

Finally, the Organizing Committee hopes that conference papers will be a good base and inspiration for all readers and will contribute to productive future work in the field of automatics, robotics and artificial intelligence.

We are looking forward to welcome you next year at 3-th edition of the conference in June, 2024.

The Committee of ICARAI'2024

IOP Conf. Series: Materials Science and Engineering **1317** (2024) 011001 doi:10.1088/1757-899X/1317/1/011001

Committee Member

Conference co-chairs:
Tsonyo Slavov - Technical University of Sofia, Bulgaria
Valeri Mladenov - Technical University of Sofia, Bulgaria

Program Committee
Kiril Alexiev - Bulgarian Academy of Science, Bulgaria
Plamen Angelov - Lancaster University United Kingdom
Pierre Borne - Ecole Centrale de Lille, France
Nikola Chalashkanov - University of Lincoln, UK
Todor Ganchev - Technical University of Varna, Bulgaria
Alexander Gegov - University of Portsmouth, United Kingdom
Petia Georgieva - University of Aveiro, Portugal
Tadeusz Kaczorek - Warsaw University of Technology, Poland
Nikola Kasabov - Auckland University of Technology, Auckland, New Zealand
Petya Koprinkova - Bulgarian Academy of Science, Bulgaria
Stefan Kozak - Slovak University of Technology in Bratislava, Slovakia
Alena Kozakova - Slovak University of Technology in Bratislava, Slovakia
Nina Nikolova - Technical University of Sofia, Bulgaria
Emil Nikolov - Technical University of Sofia, Bulgaria
Nikos Mastorakis - Hellenic Naval Academy, Greece
Utkal Mehta - University of South Pacific, Fiji
Lyudmila Mihaylova - University of Sheffield, UK
Maria Isabella Garcia Planas - Universitat Politecnica de Catalunya, Barcelona, Spain
Kamen Perev - Technical University of Sofia, Bulgaria
Aleksandar Efremov - Technical University of Sofia, Bulgaria
Petko Petkov - Bulgarian Academy of Science, Bulgaria
Danica Rosinova - Slovak University of Technology in Bratislava, Slovakia
Nikolay Stoyanov - Technical University of Sofia, Bulgaria
George Vachtsevanos - Georgia Institute of Technology, Atlanta, Georgia, USA

Organizing Committee
Nikolay Stoyanov – Technical University of Sofia
Vladimir Hristov - Technical University of Sofia
Nikolina Petkova - Technical University of Sofia
Antonia Pandelova – Technical University of Sofia
Krasimir Galabov - Technical University of Sofia
Aleksandar Ishtev - Technical University of Sofia
Boris Kirov - Technical University of Sofia, Bulgaria

Technical Committee
Antonia Pandelova - Technical University of Sofia
Alexander Marinchev - Technical University of Sofia
Bozhidar Rakov - Technical University of Sofia

IOP Conf. Series: Materials Science and Engineering **1317** (2024) 011001 doi:10.1088/1757-899X/1317/1/011001

Maya Stoycheva - Technical University of Sofia
Sonya Asenova - Technical University of Sofia

IOP Conf. Series: Materials Science and Engineering **1317** (2024) 011002 doi:10.1088/1757-899X/1317/1/011002

Peer Review Statement

All papers published in this volume have been reviewed through processes administered by the Editors. Reviews were conducted by expert referees to the professional and scientific standards expected of a proceedings journal published by IOP Publishing.

- **Type of peer review:** Single Anonymous
- **Conference submission management system:** Morressier
- **Number of submissions received:** 26
- **Number of submissions sent for review:** 26
- **Number of submissions accepted:** 18
- **Acceptance Rate (Submissions Accepted / Submissions Received × 100):** 69.2
- **Average number of reviews per paper:** 2
- **Total number of reviewers involved:** 39
- **Contact person for queries:**
  Name: Nikolay Stoyanov
  Email: n_stoyanov@tu-sofia.bg
  Affiliation: Technical University of Sofia, Faculty of Automatics

IOP Conf. Series: Materials Science and Engineering **1317** (2024) 012001 doi:10.1088/1757-899X/1317/1/012001

Robust PID control of a two-rotor aerodynamical system with decoupling

Asparuh Markovski[1*]

[1] Department of Automatics, Technical University of Sofia, Bulgaria
*E-mail: agm@tu-sofia.bg

Abstract. In this work, robust PID design via μ synthesis for controlling a MIMO two-rotor aerodynamic system manufactured by the Polish company Inteco is presented. The controller consists of a decoupling system and two PID controllers. The aim is to demonstrate an approach enabling further fine-tuning based on an already synthesized controller structure. Tools from Robust Control Toolbox in MATLAB are used. The PIDs are constructed as tunablePID objects, and the decoupler is also tunable and has the preset structure of a static gain matrix (tunableGain) or dynamic state-space system with a given number of states (tunableSS). The model contains some parameters with structured uncertainty. Several variants of the controller with a fixed structure have been created, and their performance has been tested by simulation and on the laboratory set-up.

1. Introduction

Before the new musyn function was introduced in Robust Control Toolbox for MATLAB 2021b [1], it was not possible to perform a fully automated synthesis of fixed structure robust controllers for MIMO systems. Let us summarize the advantages and disadvantages of the classical robust control. The unstructured H_∞ and μ controllers work with MIMO systems by default; they implement output feedback and do not need additional filters, state observers, scaling systems, decouplers, etc.; the control objectives can be given in the frequency domain all at once using penalty filters; their implementation in the industrial controllers is not difficult. As a drawback, they cannot be tuned additionally by hand due to their state space representation. They need the model of the object with the uncertainty parameters or dynamics to be built. Moreover, the μ synthesis procedure is a very computationally intensive task and is very difficult to be realized in real time. Robust controllers are conservative; but this is not a defect, rather a feature, because they are built for the worst possible model dynamics expected.

On the other hand, the classical PID controllers are very intuitive to be adjusted in most cases, even without a model and by hand; their structure is simple – they have three main parameters (plus the time constant of the differential part filter). But, in order to work in MIMO systems, they need additional schemes, like decouplers and filters, which are also part of the controller structure.

It is very tempting to combine the two approaches in order to make a structure that is intrinsically robust (with respect to the model uncertainty), but can then be fine-tuned by hand to make it optimal for the specific control object realization. For the structural robust synthesis, the musyn function from MATLAB can be used. It utilizes the approach described in [2]; the problem of minimizing the μ norm of the closed-loop system is solved using outer relaxation techniques via multiplying the original state space representation by scaling systems with a specific structure. Thus, the subgradient of the task acquires a special form and the steepest descent algorithm can be easily applied. Of course, the problem is non-convex and the convergence is not guaranteed, but the tuned controller structure can be used as a starting point for further work.

2. Laboratory set-up

The two rotor aerodynamic system [3], produced by the Polish company Inteco is a suitable "toy" for demonstrating the effects described above. Its view is shown in Figure 1.

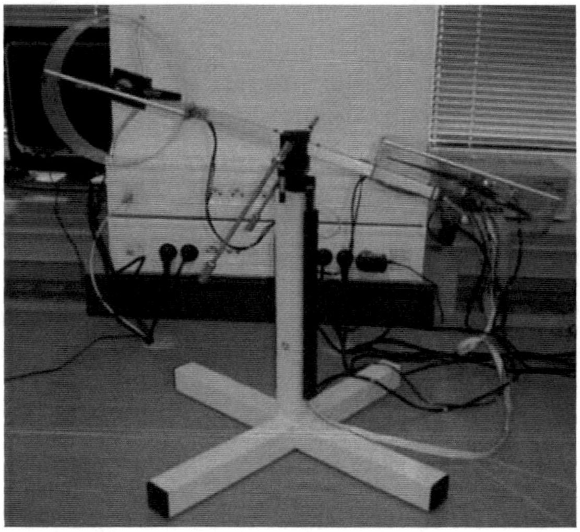

Figure 1. Two rotor aerodynamic system

It has two degrees of freedom (azimuth and pitch rotation) and two control inputs – the thrust of the main and the tail propeller. The process variables are the azimuth and pitch angles. The propellers are driven by DC motors, controlled via PWM and the whole system has an interface to MATLAB, making it a convenient tool for rapid testing of different control laws.

There is a Simulink model of the nonlinear system, given by the producer [3]. Its linearization and modelling of the uncertainty are thoroughly described in [4], so they will not be discussed here. There are 10 parameters of the linearized model with a relative degree of uncertainty from 5 to 10% from their nominal values. The model has 6 states, it has both dynamics and uncertain real parameters (of class ureal), and so it is an uss (state space with uncertainty) MATLAB model.

3. Robust synthesis of a Decoupler and two PID controllers

The control law of this set-up can be synthesized in different ways. For example, there is a cross-coupled system of 4 PID regulators, given by the producer in the form of a Simulink model. A classical (unstructured) μ controller is synthesized in [4]. Model predictive control using state space neural networks with delays is proposed in [5], and fuzzy logic is applied in [6].

There are two control channels in the set-up, so our task is to demonstrate simultaneous μ synthesis of a decoupling system and two PID controllers. The block-scheme of the open-loop connection is shown in Figure 2, and the corresponding closed-loop system, where the decoupler, PID1 and PID2 are to be tuned, is shown in Figure 3.

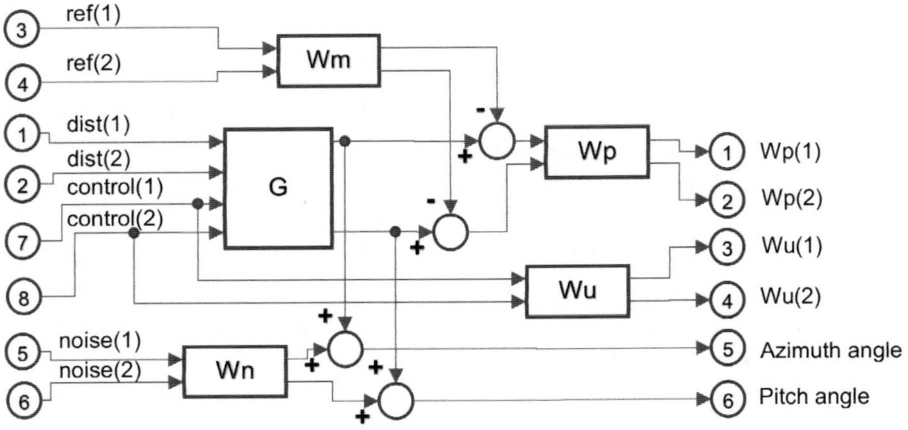

Figure 2. Open-loop interconnection

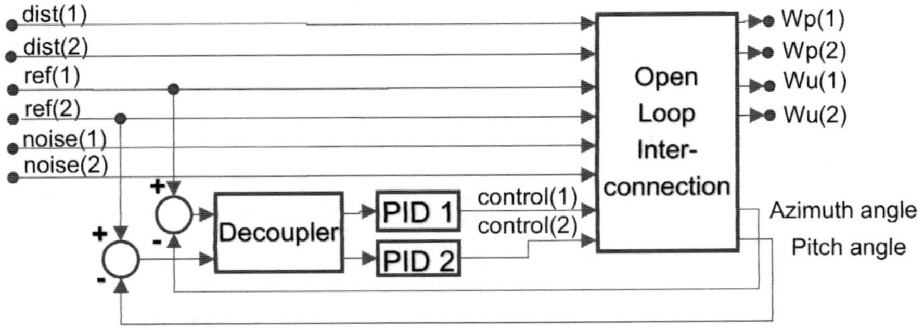

Figure 3. Closed loop system for tuning

There are dynamics, uncertainty and tunable parameters in the open-loop and closed-loop interconnection, so they are presented with genss MATLAB models. Here the load disturbances of the corresponding channels are denoted by $dist(1)$ and $dist(2)$, $ref(1)$ and $ref(2)$ are the desired azimuth and pitch angles, $noise(1)$ and $noise(2)$ are the corresponding unshaped sensor noises (the sensors are optical incremental encoders).

The uncertain model G and the corresponding penalty and shaping filters for the μ synthesis are all described in discrete time with $Ts = 0.01$ s, like in the physical set-up. The filters used are as follows: W_m is the reference model, W_p is the performance penalty, W_u is the control penalty, W_n models the sensors noise. Their values (in continuous time) are:

$$W_m = \begin{bmatrix} \frac{1}{2s+1} & 0 \\ 0 & \frac{1}{1.4s+1} \end{bmatrix}, \quad W_p = \begin{bmatrix} \frac{0.15s+3}{14s+1} & \frac{0.2}{s+1} \\ \frac{0.2}{s+1} & \frac{0.4s+40}{15s+1} \end{bmatrix},$$

$$W_u = \begin{bmatrix} \frac{2\times10^{-6}s+4\times10^{-5}}{0.0001s+1} & 0 \\ 0 & \frac{2.3\times10^{-5}s+0.0002}{0.0001s+1} \end{bmatrix}, \quad W_n = \begin{bmatrix} \frac{0.01s}{s+1} & 0 \\ 0 & \frac{0.01s}{s+1} \end{bmatrix}$$

IOP Conf. Series: Materials Science and Engineering **1317** (2024) 012001 doi:10.1088/1757-899X/1317/1/012001

3.1 μ synthesis with a static decoupler

In the first variant of the μ synthesis described here, the decoupler system is a tunable static gain with 2 inputs and 2 outputs and it is presented with tunableGain MATLAB class. The discrete tunablePID objects have the structure $W(z) = K_P + K_I \frac{T_s}{z-1} + K_D \frac{1}{T_F + T_s/(z-1)}$ and 4 tunable parameters. Only the $ref(1)$ and $ref(2)$ inputs of the closed-loop interconnection are used, for the following reasons: the bandwith of the connection from $dist(1)$ and $dist(2)$ to the penalized outputs is comparable to this from the reference inputs to the penalized outputs, but, unlike the case of full-order μ controller in [4], here the controller structure is too simple and does not allow the synthesis procedure to reach robust performance. The level of the sensor noise, described by W_n is too low and is of high-frequency type and its presence does not practically affect the behavior of the procedure. After the musyn function is applied, the robust performance is nearly reached, as seen from Figure 4, and the robust stability goal is exceeded by about 14 dB.

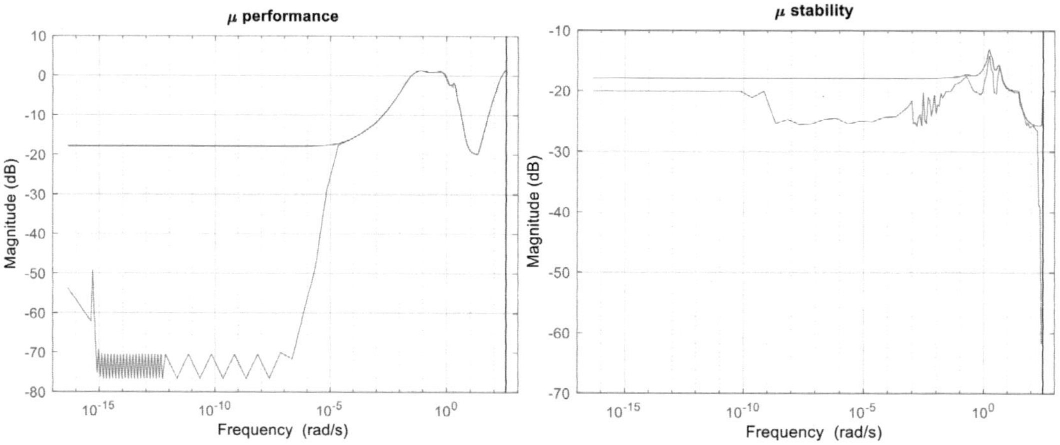

Figure 4. Robust performance (left) and robust stability (right), first variant (blue – upper μ bound, red – lower μ bound)

Simulation results with the nonlinear model for the azimuth channel are shown in Figure 5 (left), and for the pitch channel – in Figure **5** (right).

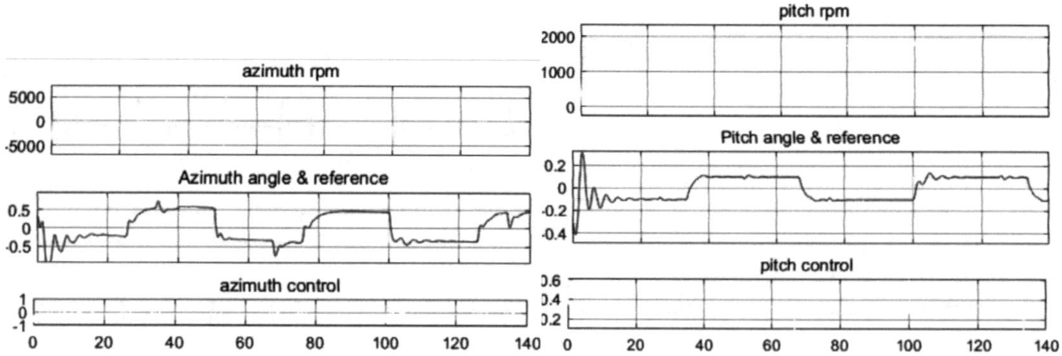

Figure 5. Simulation of the azimuth and pitch channels, first variant

The work of the physical model (without load disturbance) is shown in Figure 6.

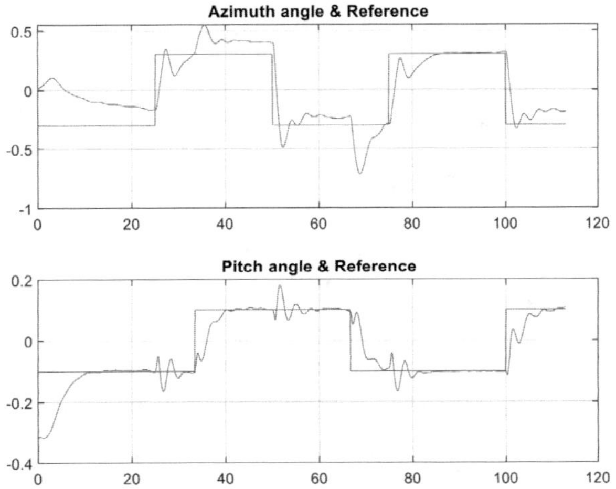

Figure 6. The work of the physical model, first variant

The values of the tuned parameters are:

Decoupler: $\begin{bmatrix} 1.16 & 1.801 \\ 0.0044 & -1.509 \end{bmatrix}$,

PID1: $K_P = 0.332, K_I = 0.0128, K_D = 0.827, T_F = 0.135$,

PID2: $K_P = -0.046, K_I = -0.165, K_D = -0.078, T_F = 0.057$

3.2 μ synthesis with a dynamic decoupler

Another approach is for the decoupler to have its own dynamics. In this second variant of the synthesis the decoupler is tunableSS object with 2 inputs, 2 outputs, 10 states and 72 tunable parameters. Its frequency response is shown on Figure 7. The tuned PIDs are:

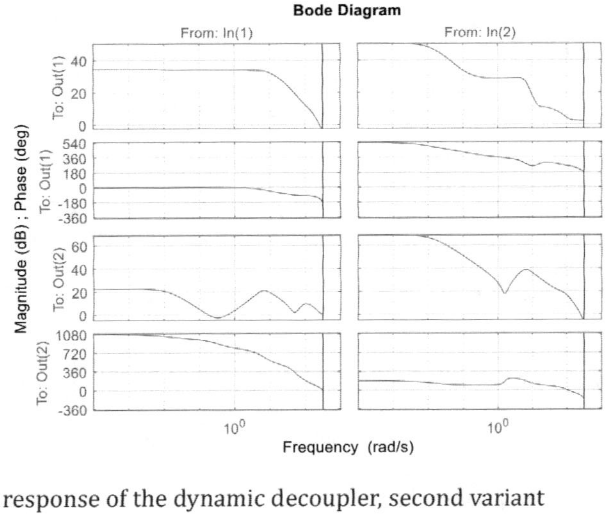

Figure 7. Frequency response of the dynamic decoupler, second variant

$$\text{PID1}: K_P = 0.0221, K_I = 0.00366, K_D = 0.025, T_F = 0.01,$$

$$\text{PID2}: K_P = -0.0073, K_I = -1.3 \times 10^{-6}, K_D = -0.0057, T_F = 0.01$$

Of course, this parameters are meaningful only combined with the decoupler. It is interesting that the performance here is not so different from that of the first variant (so it is not shown), but the disturbance rejection is better.

4. Results and discussion

The performance of the systems, described in the previous sections is not the best for the specific model realization. The mutual influence between the channels is particularly clearly visible, due to the fact that each propeller, when changing its angular velocity, creates a torque on the arm, which disturbs the other channel. The performance with the full order unstructured μ controller in [4] is definitely better.

The main advantage of this approach can be considered the obtainung of the decoupling system, which is a starting point for fine tuning (by hand, or by other methods) of the two PID regulators. This can be especially useful for more complex systems, where the construction of the decoupling system is not an easy task. Of course, the approach has its own drawbacks, mostly due to the fact that the problem for μ tuning is not convex and depends on the initial conditions of the procedure.

Other variants of the closed-loop system can be tried. For example, it is possible to include an additional aperiodic filter with a tunable time-constant (of MATLAB class realp) in the azimuth and pitch channel for pre-filtering the sensor noise. Such a filter can be then also retuned by hand, if needed.

5. Acknowledgement

This study is financed by the European UnionNextGenerationEU through the National Recovery and Resilience Plan of the Republic of Bulgaria, project №BG-RRP-2.004-0005.

References

[1] Robust Control Toolbox – documentation, Mathworks, https://www.mathworks.com/help/robust/
[2] Aguiar R., P. Apkarian, D. Noll, Structured Robust Control Against Mixed Uncertainty, IEEE Transactions on control systems technology, N 2017
[3] Two rotor aerodynamical system, Inteco, Poland, https://www.inteco.com.pl/products/two-rotor-aerodynamical-system/
[4] Petkov P., N. Christov, M. Konstantinov, Robust Real-Time Control of a Two-Rotor Aerodynamic System, Proc. of the 17th World Congress of the International Federation of Automatic Control Seoul, Korea, July 6-11, 2008
[5] Czaikowsky A., K. Patan, Model Predictive Control of the Two Rotor Aero-Dynamical System Using State Space Neural Networks with Delays, Intelligent Systems in Technical and Medical Diagnostics, Advances in Intelligent Systems and Computing, vol 230, Springer, Berlin, Heidelberg, 2014
[6] Muñoz V., Fuzzy Logic Based Control of a Two Rotor Aero-Dynamical System, Tallinn University of Technology, Faculty of Information Technology, Department of Computer Control – bachelor thesis, 2014

International Conference Automatics, Robotics and Artificial Intelligence IOP Publishing
IOP Conf. Series: Materials Science and Engineering **1317** (2024) 012002 doi:10.1088/1757-899X/1317/1/012002

Feedback control for plants with time delay

Kamen Perev[1]

[1] Systems and Control Department, Faculty of Automatics
Technical University of Sofia, Sofia, Bulgaria

*E-mail: kperev@tu-sofia.bg

Abstract. The paper considers the problem of feedback control design for plants with time delay. The role of the feedback is presented showing its ability to restrict the uncertainty in the plant model. The conditional feedback control structure is introduced and its properties for restricting model uncertainty are discussed. The presence of local feedback makes the conditional feedback structure pertinent for its usage when the plant contains a time delay element. The time delay model is replaced by its Pade series approximation. The key idea is to consider the time delay model as perturbation around its Pade series approximation. This fact allows utilizing the rational function expression in the procedure of controller design, thus avoiding using the irrational function of the time delay for this purpose. It is shown that the conditional feedback control structure gives an alternative than the Smith predictor method for controller design of plants with time delay $e^{-\theta s}$ in their structure. The efficiency of the presented method is demonstrated by a numerical example.

1. Introduction

The presence of pure time delay in the system plant model complicates the controller design procedure, especially in the case when the goal is to guarantee good tracking ability at the output and effective disturbance rejection. The main problem is that the pure time delay element has an infinite dimensional system model and therefore, it is impossible to use it directly for design of rational transfer function controller. One of the most popular and effective control structure to overcome the time delay element in process control is the structure of the Smith predictor [10]. The Smith predictor uses a local feedback connection around the controller, which contains the plant model with the time delay. In this sense, the design of the controller is performed without using the time delay element model. The main disadvantage of the Smith predictor is the inability to supress constant disturbances at the input of plants containing integrators in their structure. Similarly, this control structure is unemployable when the open-loop plant is unstable [3]. In order to overcome these difficulties, in the control literature appear different modification of the Smith predictor, see [7], [3]. In [4] a modification of Smith predictor is presented, which significantly increases the speed of reaction. The improved performance is achieved by using additional regulator that leads to two degrees of freedom control scheme. Additional regulator is also used in [11] that compensates the disturbance at the plant input. In this way the system reaction to reference and disturbance signals are separated and the regulators tuning is independently performed. The introduction of second regulator for a two degree of freedom scheme is proposed in [12], which is based on the disturbance observer approach. A simple dead time integrative for integrative processes based on disturbance estimation is proposed in [5], where robustness of the control system is guaranteed by using the stability margins for regulator design. This type of controller design is relevant not only to stable but also to unstable plants.

This paper considers the problem of control system design in presence of uncertainty, where the plant model includes time delay and integrator in its description. The main idea of the paper is to use the fundamental role of feedback in reducing the uncertainty, disturbance rejection and unmodeled dynamics in system description. A special type of control system structure is presented and its important properties are discussed. Based on these properties the controller design procedure is developed, which uses rational function approximation of the time delay element for controller design. In this way, the infinite dimensional model of the time delay model is avoided in the procedure of controller design. Based on the special system properties of control structure, which basic element is a local feedback, the algorithm for controller parameters computation is considerably simplified. A numerical example is presented, confirming its good compensation properties concerning plants with time delay and integrator element in its description.

2. The conditional feedback control structure

We consider the feedback system $\Sigma(G, K)$ shown in fig.1,

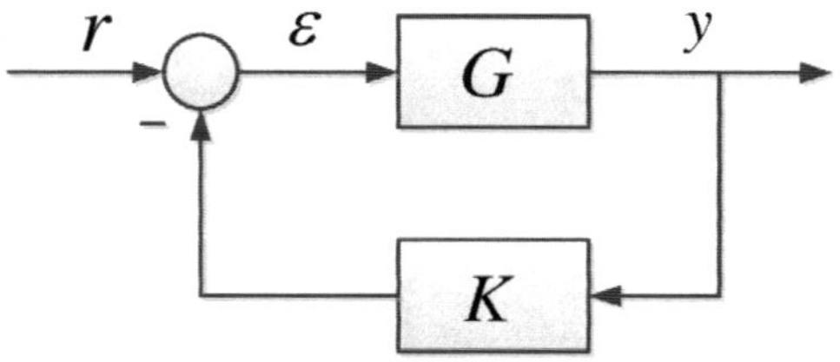

Figure 1. Standard feedback control structure

where $G(s)$ is the transfer function of the plant and $K(s)$ is the transfer function of the feedback element. The transfer function of the closed-loop system is given by the expression as follows: $H(s) = G(s)[1 + G(s)K(s)]^{-1}$. We can assume that, the closed-loop system $\Sigma(G, K)$ is equivalent to an open-loop system consisting of a precompensator $F \triangleq [1 + G(s)K(s)]^{-1}$ followed by the plant G, see [2]. The open-loop system F followed by G has the same input/output map H as $\Sigma(G, K)$. Consider now a change in the plant model $\tilde{G} = G + \Delta G$. The input/output map variation of the equivalent open-loop system is given by the following expression:

$$\Delta G \cdot F = \Delta G[1 + G(s)K(s)]^{-1} = \Delta H^0, \tag{1}$$

where the superscript of ΔH^0 implies that the change is in the equivalent open-loop system. Consider now the plant variation in the closed-loop system $\Sigma(G + \Delta G, K)$, which can be written in the form [2]:

$$\Delta H = \left(1 + \tilde{G}K\right)^{-1}\tilde{G} - G(1 - GK)^{-1} = \left(1 + \tilde{G}K\right)^{-1}\Delta G(1 + GK)^{-1} \tag{2}$$

From (1) and (2) follows that:

$$\Delta H(s) = \left[1 + \tilde{G}(s)K(s)\right]^{-1}\Delta H^0(s) \tag{3}$$

Expression (3) gives the relation between the variation of the closed-loop system and the corresponding equivalent open-loop system due to change in the plant model. The variation of the closed-loop system is equal to the variation of the equivalent open-loop system

IOP Conf. Series: Materials Science and Engineering **1317** (2024) 012002 doi:10.1088/1757-899X/1317/1/012002

premultiplied by the sensitivity function $S(s) = \left[1 + \tilde{G}(s)K(s)\right]^{-1}$. If we consider the frequency response of the system, for fixed values of ω, the relation (3) expressed as:

$$|\Delta H(j\omega)| = \left|1 + \tilde{G}(j\omega)K(j\omega)\right|^{-1}|\Delta H^0(j\omega)| \tag{4}$$

Therefore, if $\tilde{G} = G + \Delta G$ is strictly proper, $\Sigma(G + \Delta G, K)$ is exponentially stable, and if over the frequency band Ω, $\left|\tilde{G}(j\omega)K(j\omega)\right| \gg 1$, then the input/output map of the closed-loop system $\Sigma(G, K)$ is much less sensitive to plant perturbations ΔG than the input/output map of the equivalent open-loop system. In other words, in order to make the closed-loop system insensitive to uncertainty in the plant model, it is necessary to increase the magnitude response of the feedback element $|K(j\omega)|$ over the frequency band of interest Ω, see [2]. If we assume that the feedback element is simply a gain K, then this result relates directly to the Black's formula generalization, see [1]. The above derivations justify the role of the feedback for quantifying the procedure of the negative feedback amplifier operation. We consider the feedback system in fig.1, where the feedback element is a gain K. The closed-loop system transfer function is given by the expression:

$$H(s) = G(s)[1 + G(s)K]^{-1} = \frac{1}{K} \cdot \frac{KG(s)}{1+KG(s)} \tag{5}$$

where in (5), we have multiplied and divided the closed-loop system transfer function by the gain K. We assume that for a given set of frequencies $\omega \in \Omega$, $|KG(j\omega)| \gg 1$ or equivalently for the same set $|1 + KG(j\omega)| \approx |KG(j\omega)|$, then from (5) follows that:

$$|H(j\omega)| \cong \frac{1}{K}. \tag{6}$$

Therefore, for high gains in the feedback, the closed-loop system behaves similarly as the inverse feedback gain.

We consider now the conditional feedback control structure, shown in fig.2:

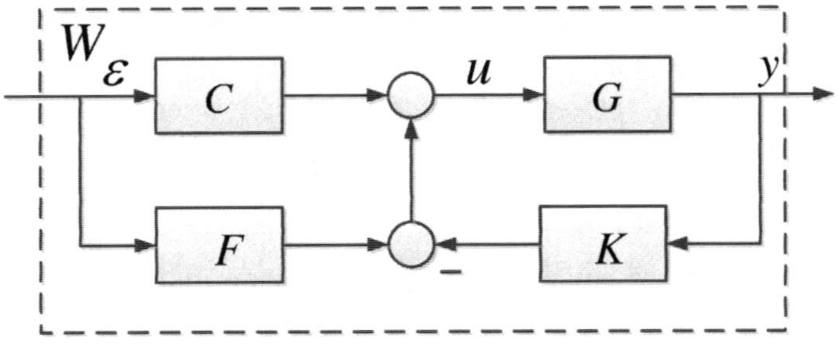

Figure 2. The conditional feedback control structure

where the transfer function $G(s)$ models the plant, $C(s)$ models the regulator, $K(s)$ models the local feedback regulator and $F(s)$ is dynamic element to be determined later. As can be seen the control structure in fig.2 represents a cascade connection of a parallel structure, consisting of the elements C and F, and a local feedback structure, consisting of the elements G and K. We assume that plant uncertainty is parametric structured uncertainty and therefore, $G(s) = G(s, \lambda)$, where λ is the uncertain parameter. We assume that the transfer function of the element, connected in parallel to the regulator C has the following transfer function:

$$F(s) = K(s)G(s, \lambda_n)C(s), \tag{7}$$

where λ_n is the nominal value of the uncertain parameter λ. The control structure in fig.2 is called conditional feedback structure, since if the plant model is derived for the nominal values

IOP Conf. Series: Materials Science and Engineering **1317** (2024) 012002 doi:10.1088/1757-899X/1317/1/012002

of the uncertain parameter, i.e. $G(s) = G(s, \lambda_n)$, then the transfer function of the whole connection becomes:

$$W(s) = [C(s) + F(s)]\frac{G(s,\lambda_n)}{1+K(s)G(s,\lambda_n)} = C(s)[1 + K(s)G(s, \lambda_n)]\frac{G(s,\lambda_n)}{1+K(s)G(s,\lambda_n)} = C(s)G(s, \lambda_n) \quad (8)$$

and the control structure in fig.2 behaves like an open-loop structure, consisting of a cascade connection between a regulator and plant. At the same time the local feedback connection in the control loop disappears. The conditional feedback control structure has some important properties, which will be discussed below. We introduce the concept of logarithmic sensitivity. The logarithmic sensitivity of a transfer function $G(s, \lambda)$ with respect to an uncertain parameter λ, is defined by the following expression:

$$LS_\lambda^G(s) = \frac{1}{G(s,\lambda_n)}\frac{\partial G}{\partial \lambda}\bigg|_{\lambda=\lambda_n} \quad (9)$$

The logarithmic sensitivity of a cascade connection of several elements with respect to uncertain parameter is the logarithmic sensitivity of the element, containing the uncertain parameter. For example for a cascade connection of three elements: $G(s, \lambda) = G_1(s)G_2(s, \lambda)G_3(s)$, the logarithmic sensitivity of G with respect to λ can be computed as follows:

$$LS_\lambda^G(s) = \frac{1}{G(s,\lambda_n)}\frac{\partial G}{\partial \lambda}\bigg|_{\lambda=\lambda_n} = \frac{1}{G_2(s,\lambda_n)}\frac{\partial G_2}{\partial \lambda}\bigg|_{\lambda=\lambda_n} = LS_\lambda^{G_2} \quad (10)$$

The logarithmic sensitivity of a feedback connection with respect to uncertain parameter in the plant is the sensitivity function, evaluated at the nominal value of the parameter, times the logarithmic sensitivity of the plant. For example, for the feedback connection:

$$H(s, \lambda) = G(s, \lambda)[1 + G(s, \lambda)K(s)]^{-1},$$

the logarithmic sensitivity of the closed-loop system with respect to the uncertain parameter λ in the plant can be computed as:

$$LS_\lambda^H = \frac{1}{H(s,\lambda_n)}\frac{\partial H}{\partial \lambda}\bigg|_{\lambda=\lambda_n} = \frac{1}{1+K(s)G(s,\lambda_n)}\frac{1}{G(s,\lambda_n)}\frac{\partial G}{\partial \lambda}\bigg|_{\lambda=\lambda_n} = \frac{1}{1+K(s)G(s,\lambda_n)}LS_\lambda^G \quad (11)$$

In other words, we derive the special case of (3) regarding the sensitivity of the feedback connection with respect to parameter uncertainty, showing that in order to make the sensitivity of the feedback connection with respect to parameter change small, we have to make the feedback gain high. Therefore, if $|K(j\omega)G(j\omega, \lambda_n)| \gg 1$ over a frequency band $\omega \in \Omega$, then the sensitivity $LS_\lambda^H(j\omega) \ll LS_\lambda^G(j\omega)$ with respect to parameter change in the plant model. If we consider the conditional feedback control structure in fig.2, we can notice that it consists of cascade connection of two parts. The first part is independent of the uncertain parameter and the second part represents a local feedback around the plant, containing the uncertain parameter. Based on the derivations above, we can write:

$$LS_\lambda^W = \frac{1}{W(s,\lambda_n)}\frac{\partial W}{\partial \lambda}\bigg|_{\lambda=\lambda_n} = \frac{1}{1+K(s)G(s,\lambda_n)}LS_\lambda^G \quad (12)$$

Therefore, due to the local feedback connection around the plant, the conditional feedback control structure has the effect of reducing the sensitivity of the system with respect to perturbations in the plant parameters if the condition $|K(j\omega)G(j\omega, \lambda_n)| > 1$ is satisfied. As larger is the feedback gain, as lower is the sensitivity to changes in the parameter values.

3. Feedback control of plants with time delay

We consider the feedback system shown in fig.3, where the transformed conditional feedback structure is located in the forward path,

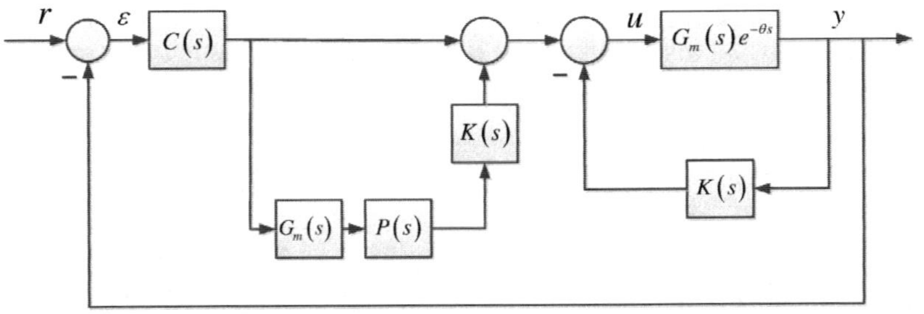

Figure 3. The closed-loop system with transformed conditional feedback in the forward path

where $C(s)$ is the transfer function of the regulator, $G(s) = G_m(s)e^{-\theta s}$ is the transfer function of the plant, consisting of minimum phase rational function $G_m(s)$ and time delay element $e^{-\theta s}$, $K(s)$ is the transfer function of the feedback element and $G_n(s) = G_m(s)P(s)$ is the transfer function of the parallel to the regulator element, where $P(s) = \frac{2-\theta s}{2+\theta s}$ is the first order Pade series approximation of the time delay element $e^{-\theta s}$.

Pade series approximation is one of the main rational function approximations for the delay element used in practice. The general form of $[n, n]$ Pade series approximation is presented by the following formula, see [8]:

$$e^{-\theta s} \approx \frac{Q_n(-s,\theta)}{Q_n(s,\theta)}, \qquad \text{with} \qquad Q_n(s,\theta) = \sum_{i=0}^{n} \binom{n}{i} \frac{\theta^i (2n-i)!}{(2n)!} s^i, \tag{13}$$

where $\binom{n}{i} = \frac{n!}{i!(n-i)!}$ is the binomial coefficient. The first several Pade series polynomials $Q_n(s)$ are given as follows: $Q_1(s,\theta) = 1 + \frac{\theta s}{2}$, $Q_2(s,\theta) = 1 + \frac{\theta s}{2} + \frac{\theta^2 s^2}{12}$, $Q_3(s,\theta) = 1 + \frac{\theta s}{2} + \frac{\theta^2 s^2}{10} + \frac{\theta^3 s^3}{120}$. The error of approximation between the $[n, n]$ Pade approximant P_n and the delay element e^{-s}, where $s = j\omega$, is bounded by the following error bound, see [8]:

$$\left| e^{-j\omega} - P_n(j\omega) \right| \le \begin{cases} 2\left(\frac{|\omega|}{\beta n}\right)^{2n+1}, & |\omega| \le \beta n \\ 2, & |\omega| \ge \beta n \end{cases}, \tag{14}$$

where $\beta = 2\left(\sqrt{2}/e\right)^{1/2} \approx 1.443$.

The key idea of the proposed control method is that the Pade series approximation is considered as the nominal model of the time delay element, while the time delay $e^{-\theta s}$ is considered as the perturbation around $P(s) = \frac{2-\theta s}{2+\theta s}$. In this sense, the nominal plant model is described by the expression:

$$G_n(s) = G_m(s)P(s) = G_m(s)\frac{2-\theta s}{2+\theta s}, \tag{15}$$

where $G_m(s)$ is a stable, minimum phase part of the plant model and $P(s) = \frac{2-\theta s}{2+\theta s}$ is an all pass system element, representing Pade series approximation of the time delay. A special role in the controller design procedure plays the feedback element with transfer function $K(s)$. Based on expression (12), it is clear that if $|K(j\omega)|$ is sufficiently high to make $|1 + G(j\omega)K(j\omega)| > 1$, then the conditional feedback structure will be less sensitive with respect to unmodeled dynamics, which in this case is represented by the time delay element perturbation $e^{-\theta s}$ with respect to the all pass element of the Pade series approximation $P(s) = \frac{2-\theta s}{2+\theta s}$. This fact will justify the usage of the Pade series approximation in the procedure of controller design. By using the properties of the conditional feedback, more particularly expression (7) with $F(s) = K(s)G_m(s)P(s)C(s)$, the usage of the transcendental irrational function for the time delay $e^{-\theta s}$ will be avoided in the

IOP Conf. Series: Materials Science and Engineering **1317** (2024) 012002 doi:10.1088/1757-899X/1317/1/012002

procedure for controller design. For example, if $K(s) = K_F$ is simply a gain with a value $K_F \gg 1$, then condition (6) will be satisfied, where the local feedback structure gain $1/K_F$ will have very small value. Then, the direct connection between the regulator and the output will be negligible, since the regulator output will be multiplied by $1/K_F$. From the other side, the connection between $G_m(s)P(s)$ and the output will be significant, since K_F and $1/K_F$ will be neutralized. In this sense, the goal to prevent using irrational terms in the controller design will be satisfied and the regulator will be determined based on the expression $C(s)G_m(s)P(s)$ in the forward path of the closed loop system. Therefore, the conditional feedback control structure gives alternative than the Smith predictor method for controller design of plants with time delay $e^{-\theta s}$ in their structure.

We consider the design strategy, where the goal is to shape directly the desired closed-loop transfer function. The presented strategy is in the form of internal model control, see [9], [6]. The main idea is to specify the desired closed-loop response and solve for the controller description. We specify the desired closed-loop transfer function from the input/output relation $y(s) = T(s)r(s)$, where $T(s)$ is the complementary sensitivity function. Based on the derivations above, we obtain the following relation:

$$T(s) = C(s)G_m(s)P(s)[1 + C(s)G_m(s)P(s)]^{-1} = f(s)P(s), \quad (16)$$

where $f(s)$ is a low-pass filter, typically of the form $f(s) = (\tau s + 1)^{-n}$ and the presence of $P(s)$ represents the influence of the time delay at the output, which cannot be prevented in the closed-loop system description. Using expression (16), we obtain the transfer function of the controller:

$$C(s) = \frac{1}{G_m(s)P(s)}\frac{T(s)}{1-T(s)} = \frac{1}{G_m(s)P(s)}\frac{1}{T^{-1}(s)-1} = \frac{1}{G_m(s)}\frac{1}{f^{-1}(s)-P(s)} \quad (17)$$

For example, if the plant minimum phase part is described by the following transfer function $G_m(s) = \frac{k}{s(Ts+1)}$ and the low-pass filter is described as $f(s) = \frac{1}{(\tau s+1)^2}$, then by using (17), the controller transfer function is obtained as follows:

$$C(s) = \frac{s(Ts+1)}{k}\frac{1}{\tau^2 s^2+2\tau s+1-\frac{2-\theta s}{2+\theta s}} = \frac{s(Ts+1)}{k}\frac{\theta s+2}{\theta\tau^2 s^3+2\tau(\tau+\theta)s^2+2(\theta+2\tau)s} = \frac{\theta T s^2+(T+\theta)s+2}{k(\theta\tau^2 s^2+2\tau(\tau+\theta)s+2(\theta+2\tau))} \quad (18)$$

4. Numerical example

We consider the control system in fig.3, where $G_m(s) = \frac{k}{s(Ts+1)}$, with $k = 2, T = 0.2sec, \theta = 0.1sec, P(s) = \frac{2-0.1s}{2+0.1s}, K(s) = K_F$. Based on (18), the controller transfer function is computed as: $C(s) = \frac{0.02s^2+0.5s+2}{0.018s^2+0.36s+2.92}$. In the following simulations, we explore the impact of the feedback gain on the system performance.

Figure 4 presents the unit step response of the closed-loop system when $K_F = 0.2$ and figure 5 presents the unit step response of the closed-loop system for $K_F = 0.2$ in the presence of constant disturbance at the plant input. It is clearly seen that the unit step response of the closed-loop system completely corresponds to the desired transfer function used in the procedure of design. The presence of disturbance at the plant input is reflected in the unit step response in figure 5. It is clear that the controller stabilizes the system response, but the speed of reaction is small and takes more than ten seconds to settle down at the final value. The system performance for $K_F = 0.2$ satisfies the design goal for system behaviour approaching that of a low-pass filter. However, the condition for reduced system sensitivity is not satisfied due to the small value of the feedback gain, i.e. $K_F < 1$. This fact explains the slow reaction of the closed-loop system to constant disturbance at the plant input.

IOP Conf. Series: Materials Science and Engineering **1317** (2024) 012002 doi:10.1088/1757-899X/1317/1/012002

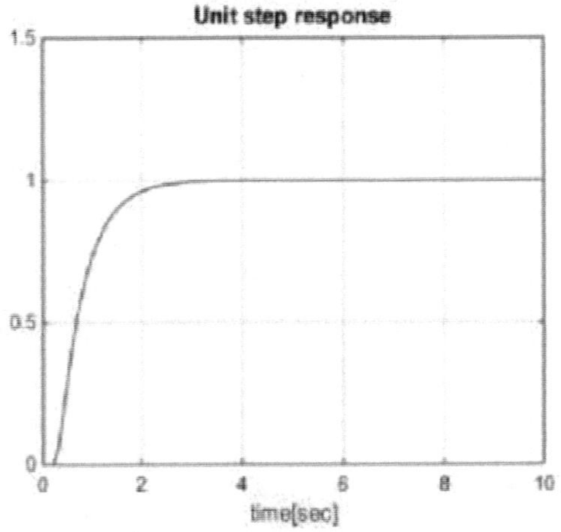

Figure 4. Unit step response $K_F = 0.2$

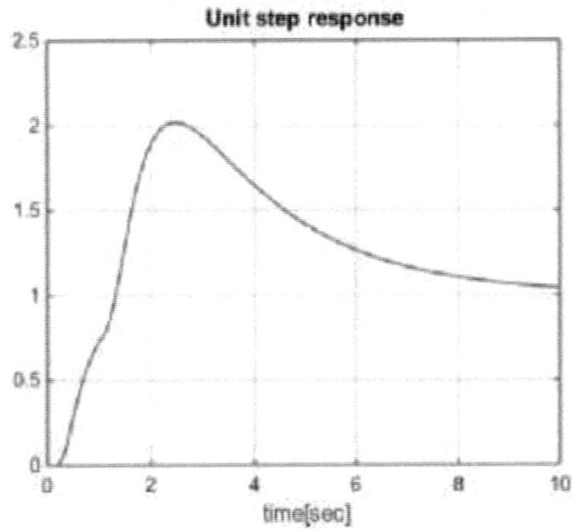

Figure 5. Step response $K_F = 0.2$ with disturbance

The time responses for higher value of the feedback gain are shown in figures 6 and 7. The unit step response of the closed-loop system for value of the feedback gain $K_F = 2$ is shown in figure 6.

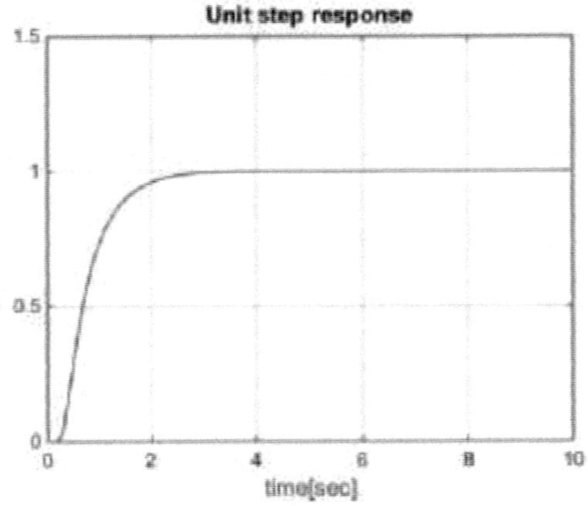

Figure 6. Unit step response $K_F = 2.0$

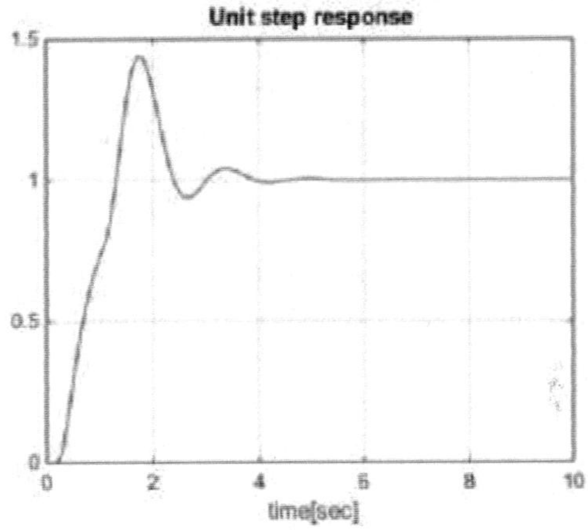

Figure 7. Step response $K_F = 2.0$ with disturbance

It is clearly seen that unit step response satisfies the design goal for the system behaviour approaching the low-pass filter behaviour. There is no considerable difference between the cases when the feedback gain is $K_F = 0.2$ and $K_F = 2$. Therefore, the delay element Pade series approximation $P(s) = \frac{2-\theta s}{2+\theta s}$ is good enough for small value of the time delay. Figure 7 represents the unit step response of the closed-loop system in presence of constant disturbance at the plant input. In comparison to the case of $K_F = 0.2$ shown in fig.5, the output reaction for $K_F = 2.0$ is much faster and the system is less sensitive to disturbances than the case for smaller feedback gains. Therefore, the conditional feedback control structure improves system performance in the cases of integrators in plant description and in presence of disturbances at the input.

Figure 8 represents the unit step response of the closed-loop system for a value of the feedback $K_F = 5.0$. The time response is similar to the time responses in fig. 4 and fig.6. Therefore, the design goal for low-pass filter behaviour is still satisfied. At the same time, a sign for aggravated relative stability is the appearance of small oscillations around the steady state value.

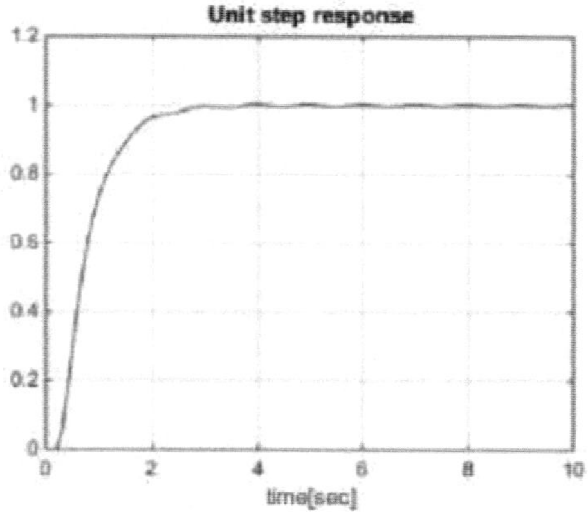

Figure 8. Unit step response $K_F = 5.0$

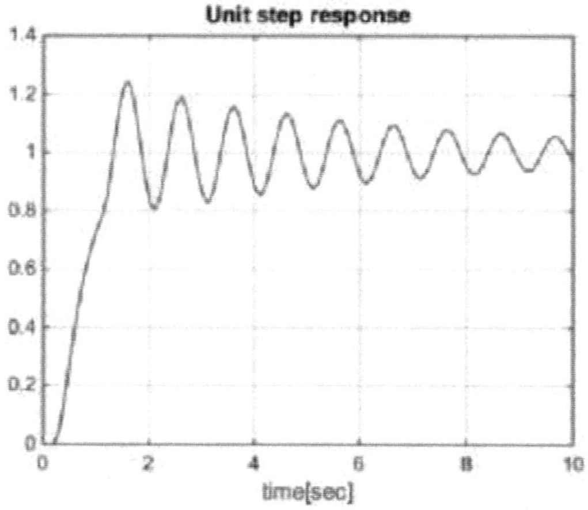

Figure 9. Step response $K_F = 5.0$ with disturbance

This fact can be clearly seen in fig.9, where the unit step response of the closed-loop system in presence of disturbance at the plant input for value of the feedback gain of $K_F = 5.0$ is shown. The oscillations around the steady-state value is a sign for low relative stability of the system. Therefore, increasing the local feedback gain reduces the system relative stability. Obviously, there exists a trade-off between reducing the sensitivity and system stability. For certain high values of the feedback gain, the closed-loop system becomes unstable. In this sense, the gain of the local feedback structure has to be carefully selected in order to preserve stability.

We consider next the case for higher value of the time delay. We assume that $\theta = 0.3sec$.

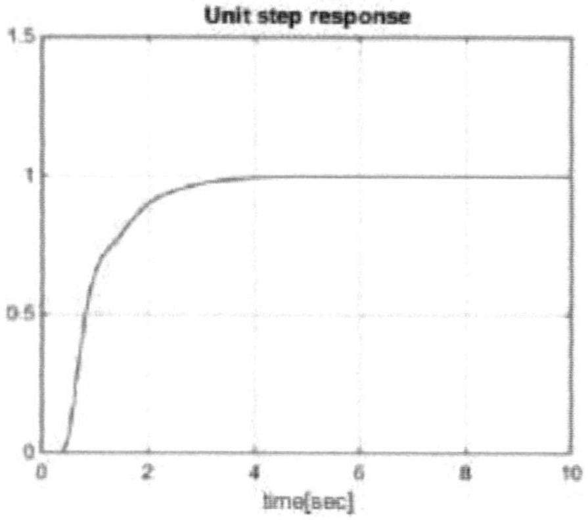

Figure 10. Unit step response $K_F = 0.2$

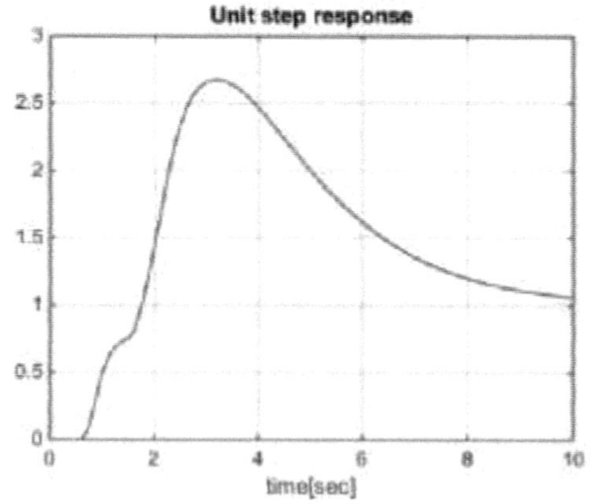

Figure 11. Step response $K_F = 0.2$ with disturbance

The unit step response for value of $K_F = 0.2$ is shown in fig.10. The system reaction is similar to the case for $\theta = 0.1sec$ except that the speed of response is slower. This is the case in presence of disturbance at the plant input, shown in fig.11. In this case, the system overshoot is slightly increased, while the time for convergence to the steady state is preserved. We can make the conclusion that when increasing the time delay, the system performance is preserved for small values of the feedback gain.

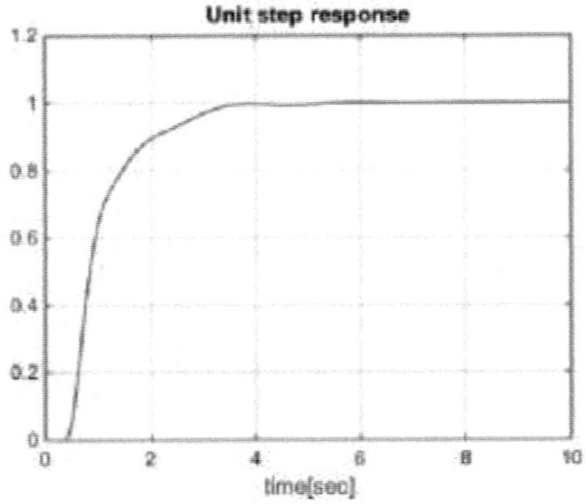

Figure 12. Unit step response for $K_F = 1.2$

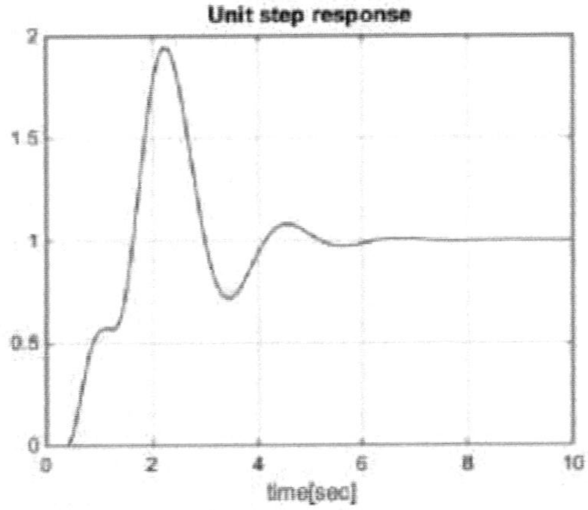

Figure 13. Unit step response for $K_F = 1.2$ with constant disturbance at the input

The unit step response for value of $K_F = 1.2$ is shown in fig.12. The settling time is slightly increased but the output reaction looks very similar to the case of smaller time delay. The unit step response in presence of constant disturbance at the plant input is shown in fig.13. The time response is much faster than the case for smaller feedback gain. The maximum overshoot is slightly increased in comparison to the case of smaller time delay as well as the settling time. In general, the system performance is almost the same, which demonstrates the good convergence properties of the conditional feedback structure with respect to plant disturbances at the input. Finally, the percentage overshoot is less than the case of smaller feedback gain, demonstrating the reduced sensitivity of the feedback structure.

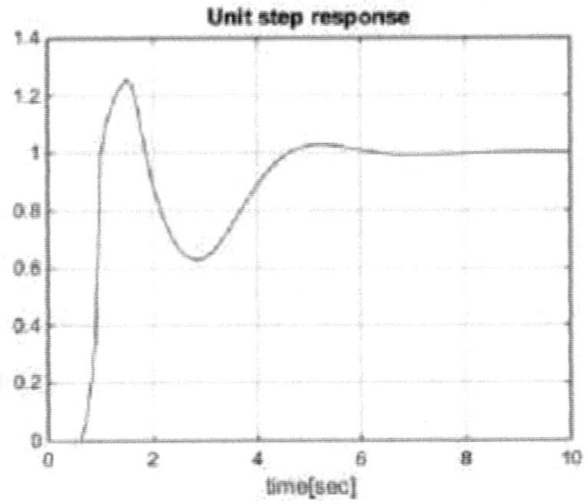

Figure 14. Step response with disturbance at the output, $K_F = 0.2$

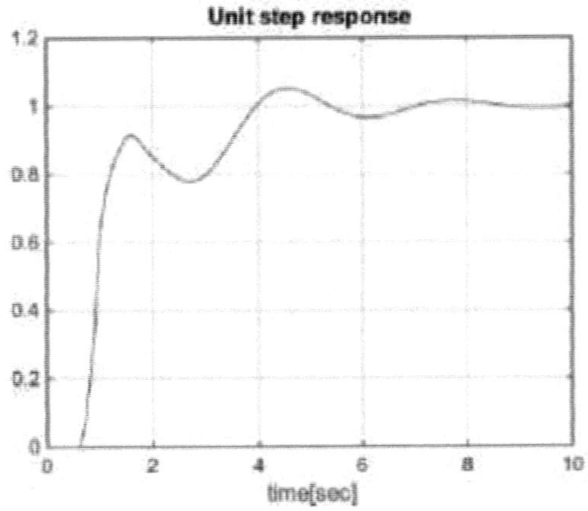

Figure 15. Step response with disturbance at the output, $K_F = 0.75$

Figures 14 and 15 show the unit step responses of the system in presence of constant disturbances at the plant output for $K_F = 0.2$ and $K_F = 0.75$. It is clear that by increasing the feedback gain, the system overshoot decreases but the settling time increases, the system becomes relatively less stable and for certain value of the gain, the system becomes unstable. Therefore, for higher value of the feedback gain, the system exhibits reduced sensitivity with respect to output disturbances, but at the same time the relative stability of the system declines.

5. Conclusion

This paper considers the problem of feedback control design for plants with time delay. The special cases of plants with integrators in their description and presence of constant disturbances at the plant input are explored. The conditional feedback control structure is introduced and its main properties are discussed. It is shown that the conditional feedback control structure has the advantage to ignore the irrational model for the time delay in the procedure of controller design. The role of the gain in the local feedback loop is explored and it is shown that under certain conditions, the conditional feedback structure has reduced sensitivity with respect to constant disturbances at the plant input. At the same time, it is shown that increasing the feedback gain above certain value leads to loss of stability. Therefore, there exists a trade-off between improved performance by reducing the sensitivity with respect to disturbances and at the same time decreasing the relative stability of the closed-loop system. A numerical example is presented to demonstrate the effects of using the conditional feedback structure to control plants with time delay in their description.

Acknowledgments

This work was supported in part by the Research and Development Sector at the Technical University of Sofia and the Systems and Control Department in the Faculty of Automatics.

References

[1] Black, H., "Inventing the negative feedback amplifier", *IEEE Spectrum*, pp. 55 – 60, 1977
[2] Callier, F., C. Desoer, *"Linear system theory"*, Springer - Verlag, New York, 1991
[3] Goodwin, G., S. Graebe and M. Saldago, *Control system design*, Prentice – Hall, Upper Saddle River, 2001
[4] Mataušek, M., A. Micić, "A modified Smith predictor for controlling a process with an integrator and long dead-time", *IEEE Trans. Autom. Control*, vol. 41, №8, pp. 1199 – 1203, 1996
[5] Mataušek, M., A. Micić, "On the modified Smith predictor for controlling a process with an integrator and long dead-time", *IEEE Trans. Autom. Control*, vol. 44, №8, pp. 1603 – 1606, 1999
[6] Morari, M., E. Zafiriou, *Robust process control*, Prentice – Hall, Englewood Cliffs, 1989
[7] Normey – Rico, J., E. Camacho, "Dead – time compensators: A survey", *Control Engineering Practice*, vol. 16, pp. 407 – 428, 2007
[8] Perev, K., "Approximation of pure time delay element by using Hankel norm and balanced realizations", *Problems of Engineering Cybernetics and Robotics*, vol. 64, pp. 24 – 37, 2011
[9] Skogestad, S., I. Postlethwaite, *Multivariable feedback control. Analysis and design*, John Wiley & Sons, Ltd. The Atrium, Southern Gate, Chichester, 2005
[10] Smith, O., "A controller to overcome dead-time", *ISA Journal*, vol. 6, №2, pp. 28 – 33, 1959
[11] Watanabe, K., M. Ito, "A process control for linear systems with delay", *IEEE Trans. Autom. Control*, vol. 26, № 6, pp. 1261 – 1269, 1981
[12] Zhang, W., H. Wang, W. Wang and X. Xu, "Novel controller for integrating and stable processes with long time delay", Proceedings of the 15th World Congress of IFAC, Barcelona, 2002

International Conference Automatics, Robotics and Artificial Intelligence IOP Publishing
IOP Conf. Series: Materials Science and Engineering **1317** (2024) 012003 doi:10.1088/1757-899X/1317/1/012003

Archaeological site detection on remote-sensed images using U-net

Adriano Gomes[1], Rodrigo Pascoal[1], Daniel Canedo[1,2], António J. R. Neves [1,2], Petia Georgieva[1,2]

[1]Dep. of Eletronics, Telecommunications and Informatics, University of Aveiro, Portugal
[2] IEETA / Department of Eletronics, Telecommunications and Informatics, University of Aveiro, Aveiro, Portugal

E-mail: petia@ua.pt

Abstract. This project aims to develop a model to identify archeological sites (tumulus and hillforts) on LiDAR data through multiclass semantic segmentation. The data used is from the Alto-Minho region in the north of Portugal and was obtained within the scope of the ODYSSEY project. In this work, it was used a U-net model, that achieved a recall and precision of 64% and 94% (class-wise metrics), respectively, regarding the identification of hillforts, and was not able to identify any of the tumulus presented on the data.

1 Introduction

This project was developed within the ODYSSEY project[1]. The main aim of the ODYSSEY project is to create a platform to aid in the process of the rigorous registration of cultural heritage. One of the main objectives of the project is to "automate the processing of the integrated information, using image processing and artificial intelligence techniques to detect archaeological sites" [1]. The use of technology, particularly Light Detection and Ranging (LiDAR) images and Artificial Intelligence, to aid in the process of discovering archaeological sites will contribute to the reduction of the difficulties felt by expert archaeologists in covering extensive and sometimes of difficult access areas.

Our aim is to explore an alternative deep learning model for archaeological site detection on remote-sensed data, particularly LiDAR derived images resulting in a Local Relief Model (LRM) of the Alto-Minho region of Portugal. The main difference of our approach to the problem was the use of a 3-class model for tumulus and hillforts detection (the third class encompasses all the rest of the landscape space, that we here on will call background).

As an alternative to the already proposed deep learning models (CNN [2], Mask-RCNN [2], YOLOv5 [2,3], YOLOv7 [4], U-net [4]) for hillforts or tumulus we propose a three-class segmentation model, based on the U-net architecture to detect both hillforts and tumulus.

[1]https://odyssey.pt/

IOP Conf. Series: Materials Science and Engineering **1317** (2024) 012003 doi:10.1088/1757-899X/1317/1/012003

Table 1: Description of the original data

Region	Resolution(px)	Annotations (tumulis)	Annotations (hillforts)
Viana do Castelo	19,978x46.000	14	41
Paredes de Coura	13,999x51,999	56	65
Arcos de Valdevez	15,999x43,999	71	46
Parque Nacional da Peneda-Gerês	19,999x44,955	135	11

2 State of the art

In the field of remote sensing (RS) archaeology, the identification of archaeological traces is an important foundation for further archaeological analysis and understanding of the ancient landscape, [5]. However, conventional methods rely heavily on visual interpretation by archaeologists and RS specialists, which is a problem due to the large amount of archaeological traces in massive RS data [5]. This drawback can be overcome through deep learning models fitted for object detection and semantic segmentation.

There are multiple researchers working with semantic segmentation and several fields where the identification of objects can be applied. Shu Yang and colleagues use an improved DeepLabv3+ model for the auto-identification of archaeological sites from very high-resolution aerial imagery. In this work, the encoder module was replaced with pre-trained ResNet101 to obtain deeper features, and, to deal with unbalanced positive and negative samples, the Dice coefficient was added to the loss function to improve the accuracy of the model [5]. The model used achieved a precision, recall, F1, and IoU of 98.37%, 98.92%, 98.64%, and 97.32%, respectively, showing the capability of this type o deep learning model in the detection of the remainings of the Great Wall in China.

Another research, conducted by Shi Yi, focuses on the semantic segmentation for Unmanned Aerial Vehicle (UAV) visual perception, referred to as crucial for UAV autonomous flight and landing spot searching. In this study, it is proposed a cascade composite transformer-based semantic segmentation network (CCTseg) for UAV images, that tries to deal with the large-scale variation of object size caused by different altitudes and angles. "The experimental results have demonstrated the effectiveness of the proposed network structure and the superiority of the proposed network over other state-of-the-art methods for the semantic segmentation of UAV visual perception" [6].

In [5] it is referred that the principal encoder-decoder network used in computer vision are the SegNet, DeepLabV3+ and the U-net. This last model is employed by [7] with model-agnostic meta-learning (MAML) and particle swarm optimization for semantic segmentation on RGB images and LiDAR data. In this case, they use the PSO to optimize the parameter updating process of the U-net MAML, improving the test accuracy. With the MAML, it was observed an increase in the accuracy of the U-net model (6%-7%) and with PSO optimization the precision increased by 6%-9%.

Within the scope of the ODYSSEY project, the capability of YoloV5, Mask R-CNN, and a custom CNN on the detection of tumulus was explored, [2]. The results were modest due to the low amount of original data and the augmented data, leading to an overfitting problem. In [3] Yolov5 was also used to identify tumulus on the LiDAR data. In his work, the main problem was the false positives, due to the fact that the model was trained with little data, becoming very sensible and detecting every shape resembling a tumulus. The precision of the model was improved using a validation process of the predictions (Local Outlier Factor algorithm).

3 Data Visualization, Description, and Pre-Processing

in the present work our starting point was a set of 4 LRM images (of the 4 regions of the Alto-Minho: Viana do Castelo, Paredes de Coura, Arcos de Valdevez and Parque Nacional da Peneda-Gerês (PNPG)) and 8 csv documents with annotations (that is the definition of MULTIPOLYGONS for each archaeological site), 4 corresponding to hillforts and 4 corresponding to tumulus, summarized in table 1.

From these 4 big images, it was possible to create a dataset of 327 images (of 640 by 640 pixels) containing at least 1 annotation; this process was employed via a script (createDataset.py, based on the platform of the ODYSSEY project). These images had 4 channels corresponding to RGB and light intensity (we kept only RGB, since light intensity was constant), for this purpose, we implement the script removeChannel.py. From these images we generated the labels for the U-net model, that we here on call masks, corresponding to images where the background was black (RGB code: 0), the hillforts are white (RGB code: 255) and the tumulus are grey (RGB code: 128). These masks were generated

Table 2: Description of the dataset

Region	Images	Total	Aug Images	Aug Total	Split
Viana do Castelo	54		324		
Paredes de Coura	103	236	618	1416	train
PNPG	79		474		
Arcos de Valdevez	91				test

using the script createUnetMasks.py. Three examples of images and their respective masks are presented below. As can be seen, there are images containing just hillforts (figures 1 and 2), just tumulus (figures 3 and 4) or both (figures 5 and 6).

With this initial set of images and corresponding masks we generated an augmented dataset, since the number of images was rather small. For this purpose, we started by selecting a test set to which we would not apply any changes (we chose the Arcos de Valdevez region since it corresponded to approximately 28% of all images, 28% of all hillforts, and 25% of all tumulus). We apply different augmentations to all images in the train set: horizontal flip, vertical flip, 90 degrees rotation, 180 degrees rotation, and 270 degrees rotation. With the original image and these results, we got an augmented train set with six times the initial length. Figures 7 to 12 illustrate the transformations applied to the train images and respective masks.

The description of the dataset is summarized in table 2, where we can se the number of images (original and augmented) for, not only train and test, but also each of the four regions.

From this train set a random train-validation split is employed before the training of the model, selecting 10% of the images for validation.

4 Short description of the U-net model

U-net is a deep learning model used for semantic segmentation, which aims at the identification of the different classes on the image. It is a pixel-wise approach, giving each pixel a classification, which differs from other models that use bonding boxes for object detection.

The architecture of the model is symmetric with a U shape and consists of two major parts, the contracting path, and the expansive path. Figure 13 shows the different blocks that are the structure of the model.

The left side of the model, the encoder, aims to obtain high-level features of the image, reducing the height and width of the inputs while the depth is increased, resulting in a compressed image and in a loss of spatial information. It follows the typical architecture of a convolutional network that consists of the repeated application of two 3x3 convolutions, each one followed by a rectified linear unit (ReLU) and a 2x2 max pooling operation with stride 2 for downsampling.

On the right side, occurs the upsampling of the feature map (decoding). In each step, the upsampling is followed by a 2x2 convolutional, a concatenation with the correspondingly cropped feature map from the contracting path (skip connections), necessary due to the loss of information in the encoding step, and two 3x3 convolutions, each one followed by a ReLU.

At the final layer, a 1x1 convolution is used to map each 64-component feature vector to the desired number of classes. In the final, we obtain a mask, with the same dimensions (width and height) as the initial image, with the different classes represented [9].

On the task of selecting the best model, it was used Cross Entropy Loss in addition to Dice Loss as the loss function, and the Dice score as a metric of the performance. within a loop, the training set was divided into several batches, used to train the model, and it was selected the one with higher validation Dice score. On the test set, the performance of the chosen model was evaluated using Recall, Precision, F1-score, IoU, and Dice score. In our case the Cross Entropy Loss is defined as:

$$CrossEntropyLoss = l(x,y) = \sum_{n=1}^{N} \frac{l_n}{\sum_{n=1}^{N} w_y y} \tag{1}$$

$$l_n = -w_y log(\frac{exp(x_n, y_n)}{\sum_{c=1}^{C} exp(x_n, c)}) \tag{2}$$

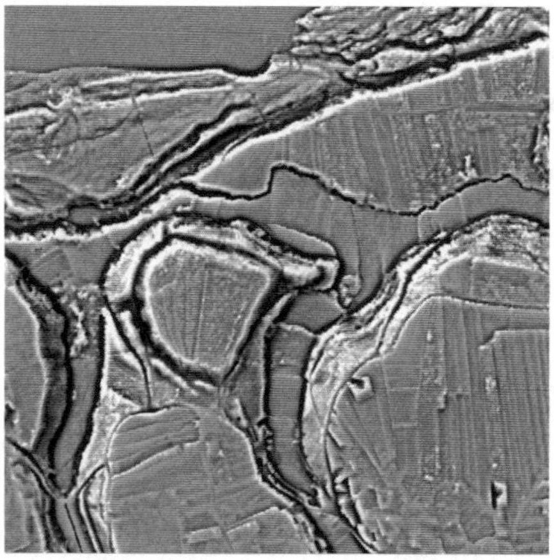

Figure 1: Hillfort

Figure 2: Hillfort Mask

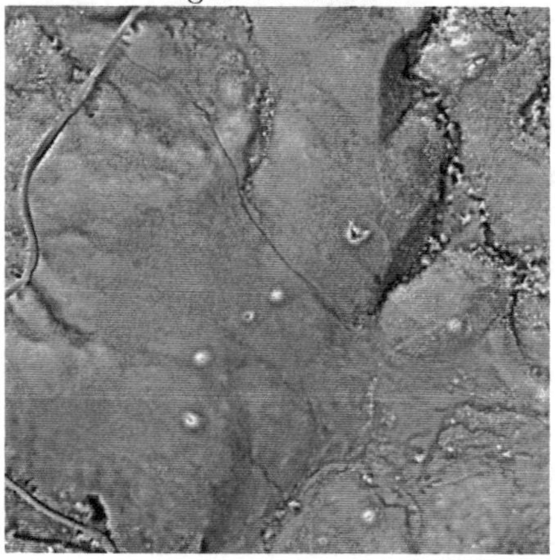

Figure 3: Tumulus

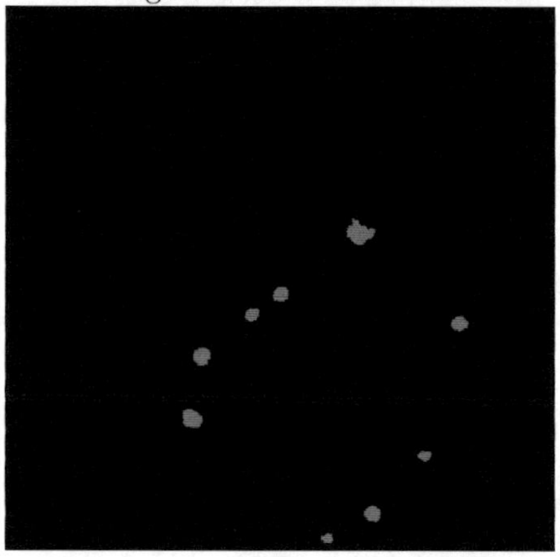

Figure 4: Tumulus Mask

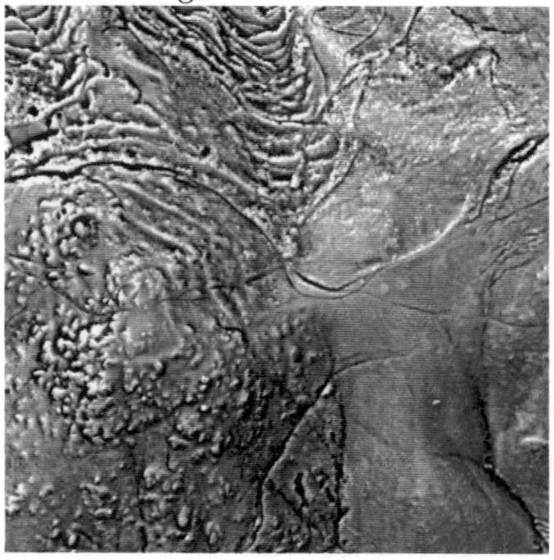

Figure 5: Hillfort and Tumulus

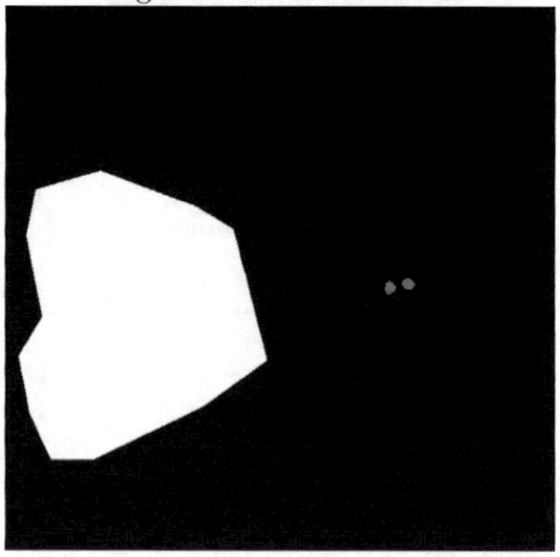

Figure 6: Hillfort and Tumulus Mask

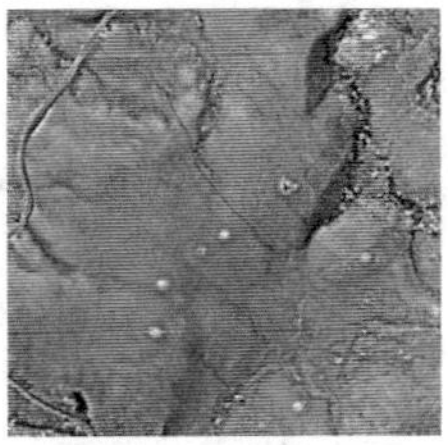

Figure 7: Original

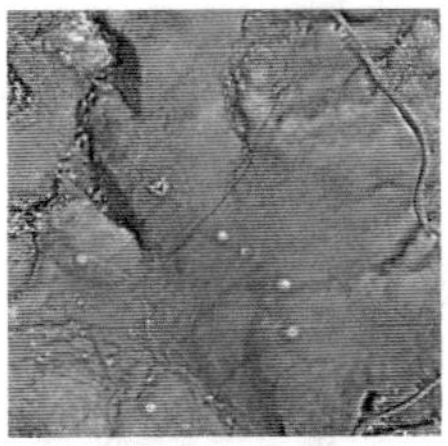

Figure 8: Horizontal Flip

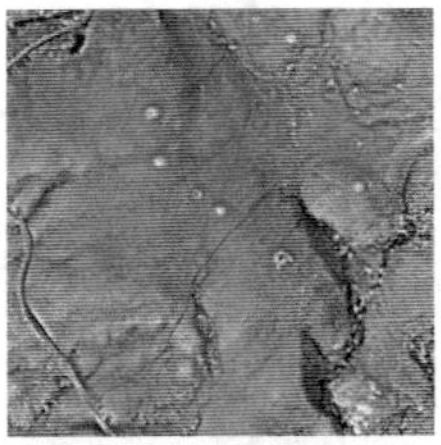

Figure 9: Vertical Flip

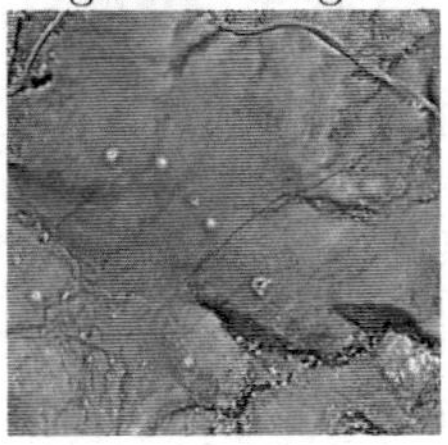

Figure 10: Rotation 90º

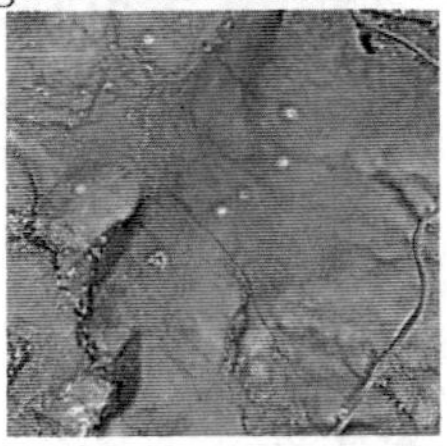

Figure 11: Rotation 180º

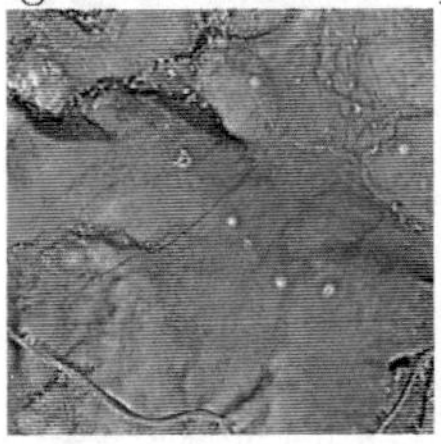

Figure 12: Rotation 270º

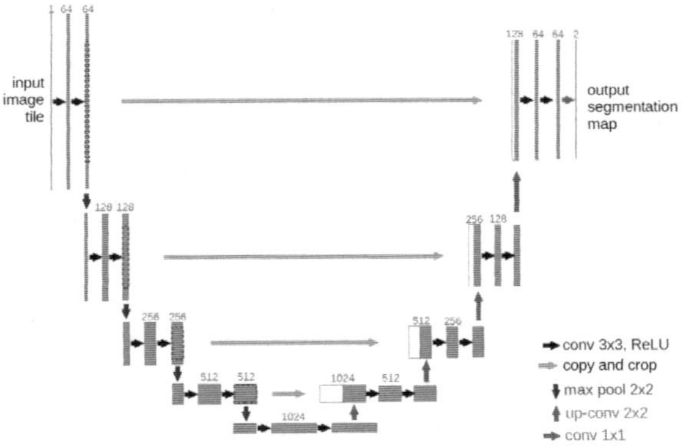

Figure 13: U-net architecture [8]

where x is the input, y is the target, w is the weights, N is the dimension of the batch and C in the number of classes.

The Dice Loss function is defined as:

$$DiceLoss = 1 - DiceScore \quad (3)$$

and Dice Score is defined as:

$$DiceScore = \frac{2 * TP}{((TP + FP) + (TP + FN))} \quad (4)$$

where TP is the True Positive (number of common pixels between prediction and groundtruth), FP is the False Positive (number of pixels incorrectly predicted as the interest class), and FN is the False Negative (number of pixels of the interest class incorrectly predicted as not the interest class). Meaning that $TP + FP$ is the number of pixels predicted in the interest class and $TP + FN$ is the number of pixels in the groundtruth of the interest class.

Intersection over Union (IoU) is calculated as

$$IoU = \frac{TP}{((TP + FP + FN))} \quad (5)$$

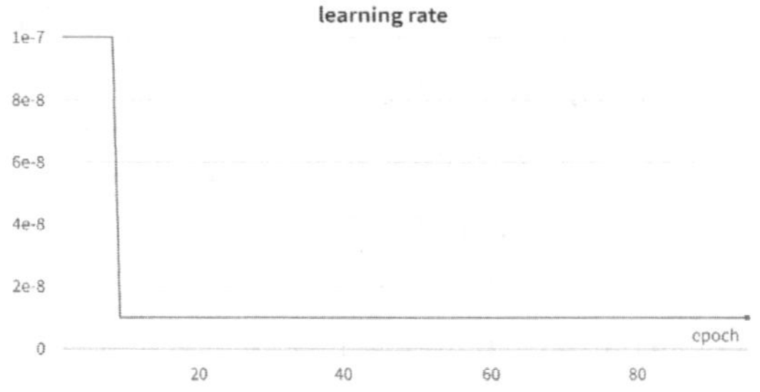

Figure 14: Evolution of learning rate by epochs

where TP is the intersection between the groundtruth and the prediction (in practice is the same as in Dice Score), $TP + FP + FN$ is the union between the groundtruth and the prediction.

Recall, Precision and f1-score are deffined as follows:

$$Recall = \frac{TP}{(TP + FN)} \tag{6}$$

$$Precision = \frac{TP}{(TP + FP)} \tag{7}$$

$$F1 = \frac{2 * Recall * Precision}{Recall + Precision} \tag{8}$$

for pixel wise performance evaluation TP, FP and FN can be defined as explained above, for class wise performance evaluation, in our work, we define TP as the number of images where an object is predicted and it exists in the grountruth, FP as the number of images where an object is predicted and it does not exist in the grountruth and FN as the number of images where an object is not predicted and it exists in the grountruth (objects:tumulus or hillforts).

In this work, it was used an adaptive learning rate in order to use the most suited value throughout the epochs, and non-other parameters were optimized since it was followed the indicated values resulting from the experience of the colleagues working on the same problem.

5 Results & Discussion

In this project, we applied the U-net architecture presented in the section above with adaptive learning rate. Starting at the value 1e−7 and changing to 1e−8 at the ninth epoch and staying at that value till the end of the training steps, as can be seen in figure 14. This indicates that the learning rate optimal values should be 1e−8.

During the training, for each epoch, the Cross Entropy Loss and the Dice index were calculated and summed to get the training loss, which is represented in figure 15. As can be seen in the five initial epochs the drop in train loss occurs sharply from a loss value of around 1.83 to 0.13; in the following epochs the loss value plateaus between 0.50 and 0.01.

The evaluation of the model trained in each epoch is done with the Dice index on the validation set. The initial value was 0.03 and in 3 epochs it raises to 0.50 where it plateaus until epoch 20. In the following epochs, it reaches a maximum of 0.57 at epoch 54 and a minimum of 0.42 at epoch 73. Since we applied early stooping in 40 epochs without improvement the training of the model stopped at epoch 95, as can be observed in figure 16.

With the best model, achieved in epoch 54, we evaluate its performance in the training and test sets. An example of an image, its ground-truth mask and the corresponding prediction can be seen in figures 17, 18 and 19. To evaluate the model we choose to focus on the interest classes (tumulus and hillforts) calculating class-wise metrics for these two classes and not for the background since it would be overly optimistic. The model was unable to predict tumulus (having value 0 for all metrics), we opted to present the metrics for the train and test sets in table 3.

All these metrics were calculated for each image by comparing the ground truth mask with the predicted label in a pixel-wise manner and averaged for all the images in the respective set. In the train

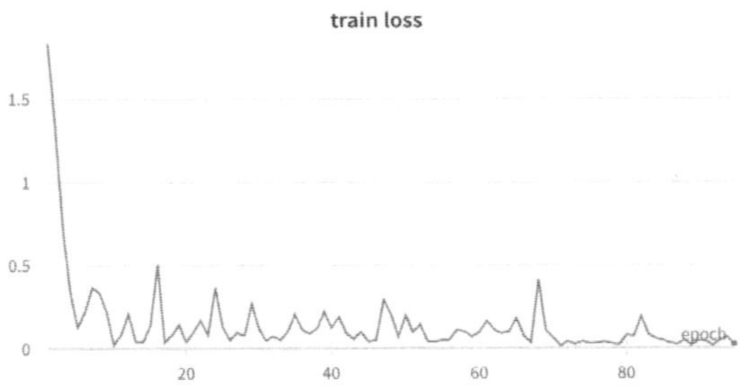

Figure 15: Evolution of train loss by epochs

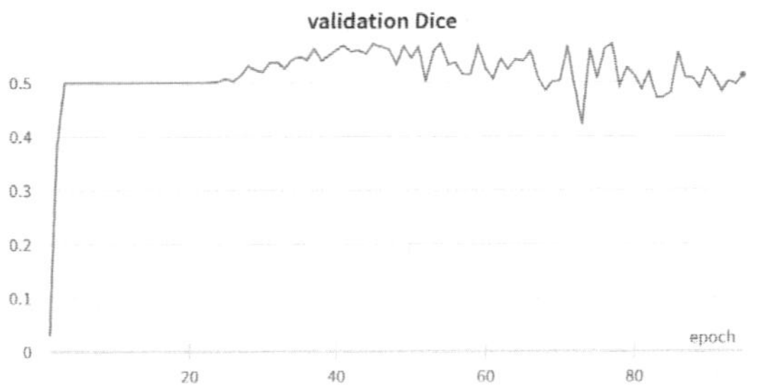

Figure 16: Evolution of validation dice index by epochs

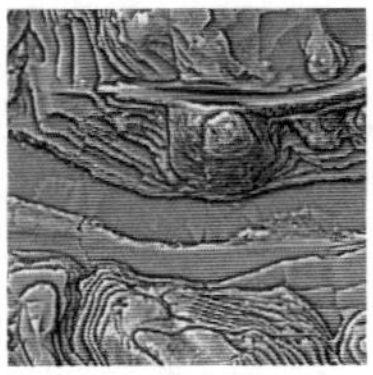

Figure 17: Original

Figure 18: Mask

Figure 19: Prediciton

Table 3: Pixel-wise performance metrics of final model on train and test sets

		Recall	Precision	F1-score	IoU	Dice Index
train set	Hillforts	0.46	0.85	0.60	0.38	0.48
	Tumulus	0.00	0.00	0.00	0.00	0.00
test set	Hilforts	0.16	0.75	0.26	0.14	0.20
	Tumulus	0.00	0.00	0.00	0.00	0.00

Table 4: Class-wise performance metrics of the final model on train and test sets

	Recall	Precision	F1-score
train set	0.94	0.92	0.93
test set	0.64	0.94	0.76

set the model correctly predicts 46% of the pixels corresponding to hillforts and 85% of the predictions are correct, that is it predicts almost half of the hillforts and it does not predict much of the background (or tumulus) as hillforts. Also, we see that from the total area of overlap between real hillforts and predicted hillforts, only 38% is correctly predicted by the model. In the test set the model correctly predicts just 16% of the pixels corresponding to hillforts and 75% of the predictions are correct, that is it predicts less than a quarter of the hillforts and a quarter of the pixels predicted as hillforts are background (or tumulus); and just 14% of the overlap area is correct.

The model appears to be slightly overfitted, possibly due to the fact that it uses the same image (only with geometric transformations) 6 times in the training stage. A possible solution would be to apply different augmentation techniques or randomly apply these transformations but not apply all of them to the same image.

Evaluating the model with class-wise metrics, now just for hillforts since we know the model was unable to predict tumulus, we get the results in table 4

Since for these metrics we assumed that whenever the model predicts a structure it is predicting it at the correct location in the image, and that there is only one object per image their values might be overly optimistic. We can see that in the train set the model correctly predicts 94% of all hillforts, that is from the 696 images with hillforts the model is able to detect hillforts in 653 images. Also, from the 710 predictions of images with hillforts 92% are correct. In the test set the model correctly predicts 64% of all hillforts, that is from the 45 images with hillforts the model is able to detect hillforts in 29 images. Also, from the 31 predictions of images with hillforts 94% are correct. Again, we can conclude that there is some overfitting in the model.

The bigger problem with our experiments is clearly the lack of power to predict tumulus, either from the data augmentation or the training of the model somethings should be changed to achieve our goal of having a single model for both classes.

The data augmentation techniques employed in this work are simpler and when compared with more complex techniques like (copyPaste as described in [4] and [3]) prove to have lower performance. Future work should explore other data augmentation techniques applied to the original dataset of this work.

The number of epochs during the training stages was much smaller than in the case of the U-net used previously as can be seen in figure 20. We employed early stopping at 40 epochs without validation dice improvement, in our experiment the highest validation dice was 0.5723 at epoch 54 and at epoch 95 we stopped training.

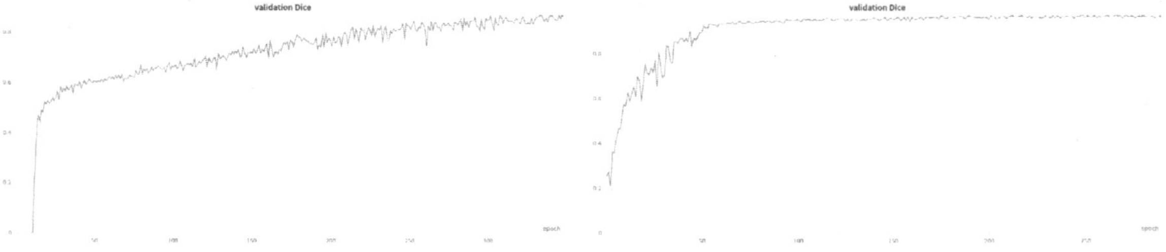

(a) Tumulus (b) Hillforts

Figure 20: Evolution of validation dice index with simple augmentation datasets for U-net models [4]

The smaller size of the tumulus, hence a smaller number of pixels, may be affecting the weights that the models used for this class, when predicting our models completely ignored the tumulus class, being unable to recognize any tumulus. This problem is clear with tumulus but also present for hillforts since most pixels in our images are background. One possible solution might be to use information about landscape occupation to create other categories, a similar approach to this is being implemented, by manually filtering results after model prediction when the problem was the high number of false positives with YOLOv5 in [3]. Alternatively, the problem might be solved with data augmentation techniques similar to copyPaste.

In [4], even with some differences in the methodology (particularly changes in test set selection) we can see that the performance of the best model (a U-net with copyPaste augmentation), when evaluated in a class-wise manner, his higher (recall), being able to correctly predict 83% of hillforts. Also, it is possible to conclude that not only U-net has higher performance when compared with YOLOv7, but that the applied augmentation technique has a high impact on performance. However, all models in [4]) suffer from the same problem, that is the low precision value indicates that there is a lot of background classified as hillforts. The proposal in [3], with higher complexity, includes filtering techniques that result in a higher precision value without reducing the recall, that is accurately predicting objects and reducing the false positives (background predicted as being an object).

Table 5: Class-wise metrics comparison with bibliography [4]

model	augmentation	Recall	Precision	F1-score
	Geometric	0.64	0.94	0.76
Unet	Transformation			
	copyPaste	0.83	0.51	0.63
	Simple	0.66	0.04	0.07
YOLOv7	Simple	0.55	0.06	0.10

6 Conclusion

The objective of this work was to create a model for semantic segmentation of LiDAR data, for the detection of archeological sites, in order to help archeologists in their work of discovering precious pieces of History. It was used the U-net model on an augmented dataset of images with objects of classes tumulus and hillforts. The achieved model was able to detect hillforts, in train, with a recall, precision, f1-score, IoU, and Dice score of 46%, 85%, 60%, 38%, and 48%, respectively (pixel-wise classification); and in test 16%, 75%, 26%, 14%, and 20%. However, with this model, we were not able to detect tumulus. It was not detected something on the algorithm that could lead to this result, maybe it was better to use a specific algorithm to identify just the tumulus class, given the small size of this type of object, in comparison with hillforts. For future work, we suggest the use of a balanced dataset to increase the performance of the model, although the use of the loss function selected to deal with unbalances. Due to the lack of images and the necessity of data augmentation, we also propose other types of transformation, like the copyPaste method, already mentioned in this work. Testing other models, such as Mask R-CNN or a variant of U-net, is also a possibility to better resolve this problem. Regarding the metric used, we suggest the exploration of methodologies and/or metrics that take into account the location of the objects and their number (problem: For an image with two objects of the same size, we can achieve a recall of 50% if the algorithm classifies all pixels of just one of the objects with the right class or if it classifies half of the pixels of each object correctly, but in the first case it just detected one object and in the second case it detected two, which is different for the sake of this application).

Acknowledgment

This work was mainly funded by the project "Odyssey: Platform for Automated Sensing in Archaeology" (Ref. ALG-01-0247-FEDER-070150), co-financed by COMPETE 2020 and Regional Operational Program Lisboa 2020, through Portugal 2020 and FEDER. The authors are grateful to the Comunidade Intermunicipal do Alto Minho (CIM Alto Minho) for providing the airborne LiDAR data used in this work. The work was supported by the European Union-NextGenerationEU, through the National Recovery and Resilience Plan of the Republic of Bulgaria, project N0 BG-RRP-2.004-0005. It was also funded by FCT/MCTES through national funds and when applicable co-funded EU funds under the project UIDB/50008/2020-UIDP/50008/2020.

References

[1] ODYSSEY - sensing project `https://odyssey.pt/` accessed: 2023-06-11

IOP Conf. Series: Materials Science and Engineering **1317** (2024) 012003 doi:10.1088/1757-899X/1317/1/012003

[2] Botelho F 2022 *Identification of archaeological sites in aerial image using Deep Learning* Master's thesis University of Aveiro URL `http://hdl.handle.net/10773/35071`

[3] Canedo D, Fonte J, Gonçalves Seco L, Vázquez M, Dias R, do Pereiro T, Hipólito J, Menéndez-Marsh F, Georgieva P and J R Neves A 2023

[4] Maia L 2023 *Archaeological site identification on aerial imagery using deep learning - ODYSSEY project* Master's thesis University of Aveiro

[5] Yang S, Luo L, Li Q, Chen Y, Wu L and Wang X 2022 *International Journal of Applied Earth Observation and Geoinformation* **113** 102995 ISSN 1569-8432 URL `https://www.sciencedirect.com/science/article/pii/S1569843222001868`

[6] Yi S, Li J, Jiang G, Liu X and Chen L 2023 *Measurement* **211** 112612 ISSN 0263-2241 URL `https://www.sciencedirect.com/science/article/pii/S0263224123001768`

[7] Zhang K, Han Y, Chen J, Zhang Z and Wang S 2020 *IFAC-PapersOnLine* **53** 397–402 ISSN 2405-8963 3rd IFAC Workshop on Cyber-Physical Human Systems CPHS 2020 URL `https://www.sciencedirect.com/science/article/pii/S2405896321002238`

[8] jeremy 2019 Unet — line by line explanation Retrieved [09-06-2023] from , `https://towardsdatascience.com/unet-line-by-line-explanation-9b191c76baf5`

[9] Ronneberger O, Fischer P and Brox T 2015 U-net: Convolutional networks for biomedical image segmentation (*Preprint* `1505.04597`)

IOP Conf. Series: Materials Science and Engineering **1317** (2024) 012004 doi:10.1088/1757-899X/1317/1/012004

Gathering information for the external parameters of Permanent Magnet Synchronous Motor using Hall sensors and applying linear interpolation

Kamen Hristov[1*]

[1] Electric Drive Automation, Technical University of Sofia, Sofia, Bulgaria

*E-mail: khristov@tu-sofia.bg

Abstract. This paper presents review of a method for receiving information for the rotational speed using the signal gathered from Hall sensors. They are widely used, because of their high performance in the whole operating range. There are shown models and substructures developments and simulation results in Matlab/Simulink.

1. Introduction.

Permanent magnet motors are widely used in electrical systems because of their high performance, such as high torque, high power and high efficiency. Hall sensors are used to estimate the rotor position, but have some disadvantages such as cost, reliability and thermal dependence [1]. Increasing the number of sensors leads to an increase in the accuracy of the rotor position estimation. In general, three Hall sensors are used for position estimation. The rotor phase estimated by the sensors has good performance in the established mode, but there are disturbances during acceleration, braking or variable movement. The method using an adaptive algorithm observer is less common in permanent magnet motor control. When receiving information about the estimated value of the phase from the observer, there are no disturbances when the speed changes, but there are disturbances in the established mode, which are caused by the algorithm of the observer itself [6]. The adaptive algorithm observer is widely used for sensorless control of permanent magnet synchronous motors because it has the properties to compensate for speed changes. As a disadvantage, it can be noted that the observer cannot drive the motor alone (without the presence of other devices) and the estimated value of the phase by the observer contains components that reduce its accuracy [2], [3]. Hybrid observer - The position of the rotor can be found from the information available from the observer. However, the use of the classic observer has two main limitations: it cannot be used in a fixed mode, at low speed and maximum load; the estimated error values may increase.

2. Phase estimation processes using Hall sensors.

3 Hall sensors are used to estimate the rotor phase. One revolution of the motor shaft contains 23 electrical periods because of the 23 pole pairs, and one electrical period is divided into six different steps by the three Hall sensors [3], [4]. The sensor signal for one period is shown in the figure.

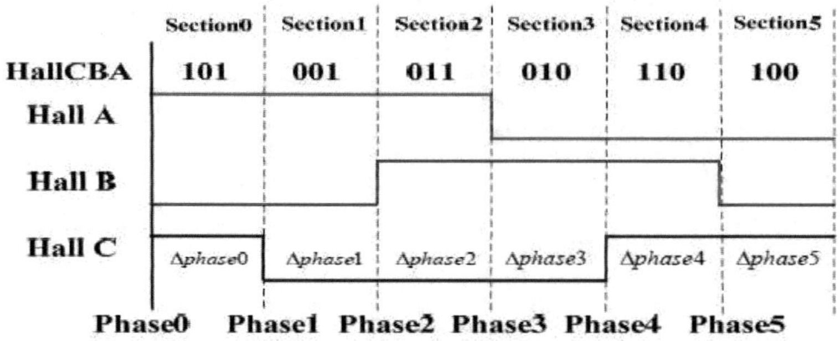

Figure 1. Three Hall signals during one electric period

The six steps can be determined by taking the values from the sensors (Hall_C, Hall_B, Hall_A) which are 5, 1, 3, 2, 6, 4. There are also 6 phase values that correspond to the values of these 6 steps, which are Phase0, Phase1, Phase2, Phase3, Phase4, Phase5.

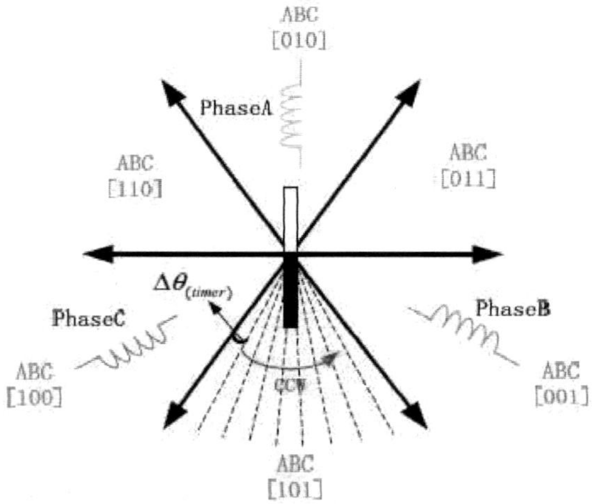

Figure 2. Phase voltages.

If the direction of rotation of the motor shaft is counterclockwise, then the values from the Hall sensors change correctly according to the set sequence. Every single change in the values from these sensors is detected by the processor (ECU) and read in order to determine which step the rotor is in. A major problem is estimating the rotor position in each of the six steps, which can be solved by using a timer. It is applicable to both simulations and real systems. The speed of rotation of the motor shaft can be determined by the formula:

$$\omega = \frac{\Delta Phase(i)}{P_n \Delta t}, i = 0,1,2,3,4,5 \quad (1)$$

Theoretically, ΔPhase(i) is 60 electrical degrees. The three sensors have an undefined error in their installation and therefore the values of ΔPhase(i) are undefined. The rotor phase increases for one period, the value is obtained from Eq.

$$\Delta\theta_{(timer)} = \frac{\omega P_n}{f_{timer}} \quad (2)$$

where Δθ(timer) is the phase gain for the particular period and ftmier is the timer frequency.

Permanent magnet synchronous motors can be driven by Hall sensors. This is essential in the practical implementation of the system. In steady state, control with Hall sensors has better performance than control with adaptive observer. However, the dependence of Hall sensors on changes in speed can be noted as a drawback.

3. Mathematical model of permanent magnet synchronous motor in d-q coordinate system.

Equations for the stator voltage projected along the d and q axes.

$$V_{qs} = R_s i_{qs} + i_{qs} pL_{qq} + L_{qq} pi_{qs} + L_{qd} pi_{ds} + i_{ds} pL_{qd} + \lambda_{af} psin\theta_r \quad (3)$$

$$V_{ds} = R_s i_{ds} + i_{qs} pL_{qd} + L_{qd} pi_{qs} + L_{dd} pi_{ds} + i_{ds} pL_{dd} + \lambda_{af} pcos\theta_r \quad (4)$$

The third term in the system is contained due to the presence of deviations (when Lq is not equal to Ld) [3], [5]. In most systems the inductances are equal and therefore L2 is equal to zero, therefore the third term is zero and does not enter the equation. Also, the second term is omitted from the system of equations, as it depends on the position.

As a final system of equations, the system independent of the rotor position and projected in the stator coordinate system can be written:

$$\begin{bmatrix} V_{qs} \\ V_{ds} \end{bmatrix} = R_s \begin{bmatrix} i_{qs} \\ i_{ds} \end{bmatrix} + \frac{\begin{bmatrix} L1 & 0 \\ 0 & L2 \end{bmatrix} d}{dt \begin{bmatrix} i_{qs} \\ i_{ds} \end{bmatrix}} + \lambda_{af} w_r \begin{bmatrix} cos\theta_r \\ -sin\theta_r \end{bmatrix} \quad (5)$$

IOP Conf. Series: Materials Science and Engineering **1317** (2024) 012004 doi:10.1088/1757-899X/1317/1/012004

4. **Application of the method with interpolation of the discretized values of the stator current.**

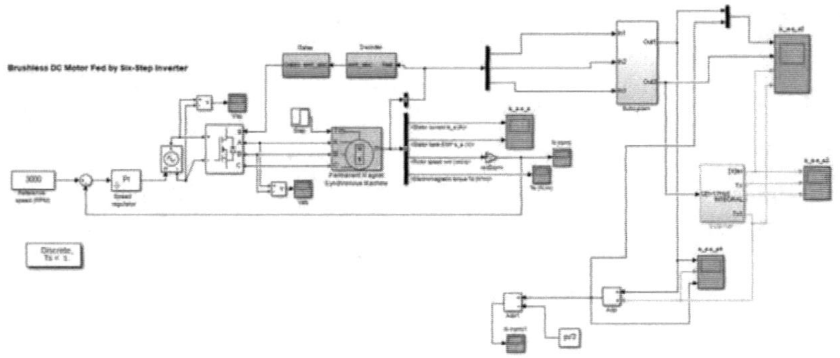

Figure 3. General scheme of the experimental setup in Simulink

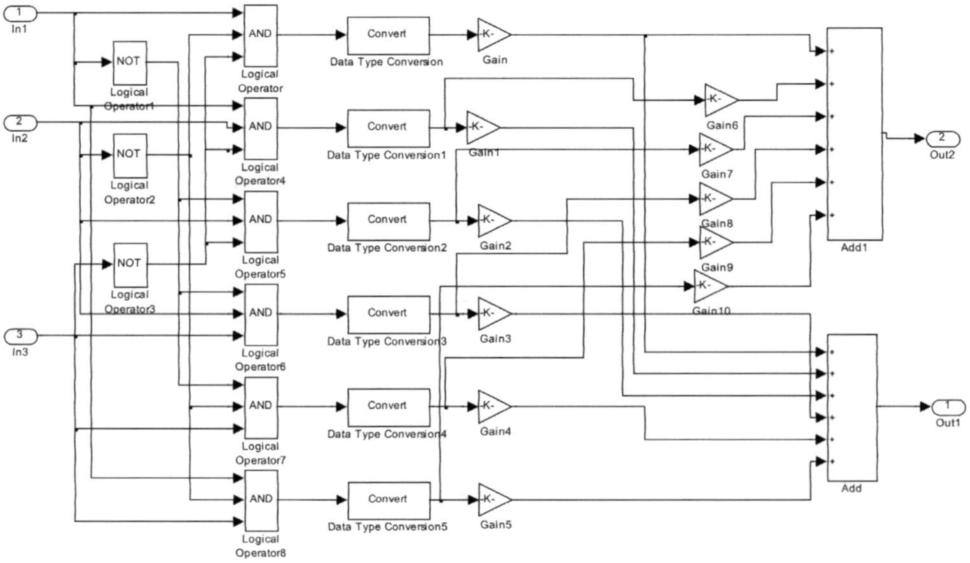

Figure 4. Logical conversion required for interpolation

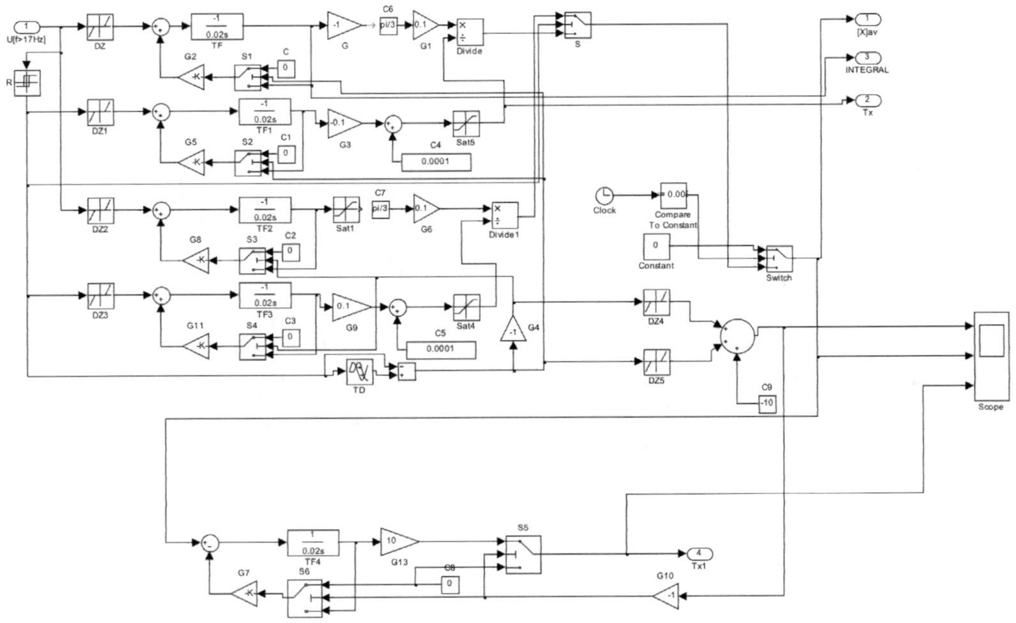

Figure 5. Subsystem Xav

5. Experimental Results

On figure 6 is shown the voltage of the DC bus, which is gathered after PI regulator and feedback of interpolation. The presented experiments are taken by estimation from Hall sensors. On figure 7 are shown voltages in phases a and b in order to clarify the correct references for the motor. Figures 8, 9 and 10 presents as follows – stator current, back EMF, motor speed and torque before interpolation. Figure 11 shows stator current, back EMF after interpolation. The main advantage of interpolation is lowered stator current which provides good energy inficators.

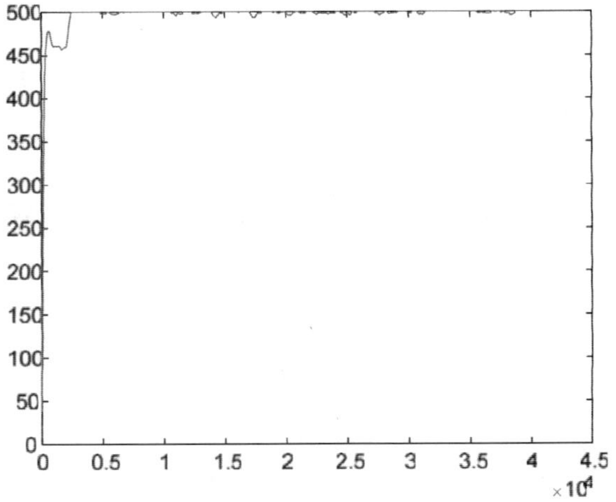

Figure 6. Voltage of the DC bus

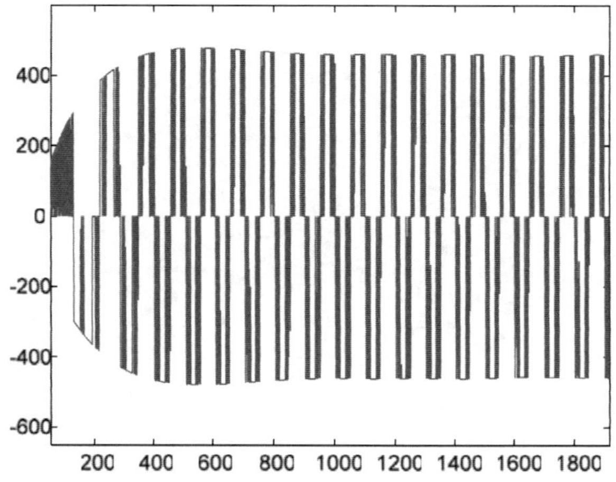

Figure 7. Voltage in phases a,b

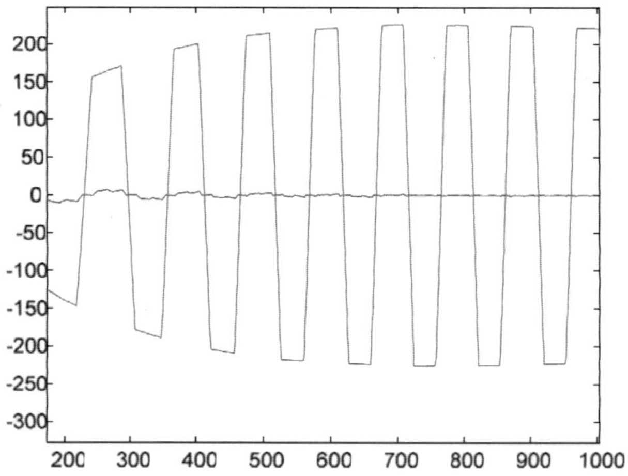

Figure 8. Stator current and Back EMF before interpolation

IOP Conf. Series: Materials Science and Engineering **1317** (2024) 012004 doi:10.1088/1757-899X/1317/1/012004

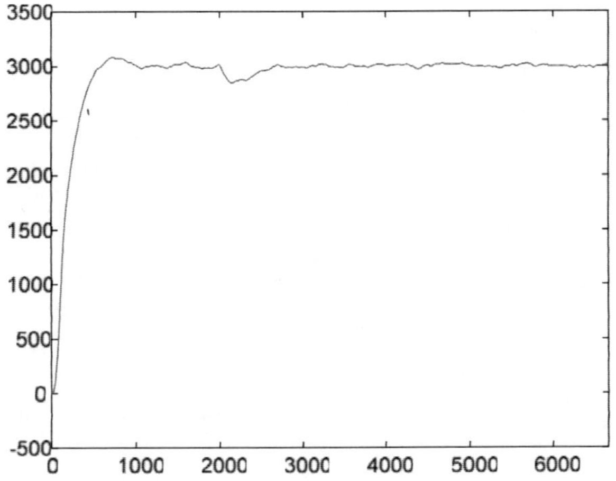

Figure 9. Motor speed before interpolation

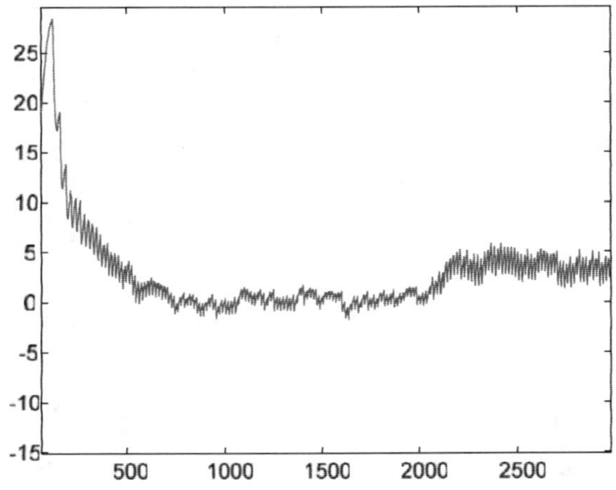

Figure 10. Motor electromagnetic torque

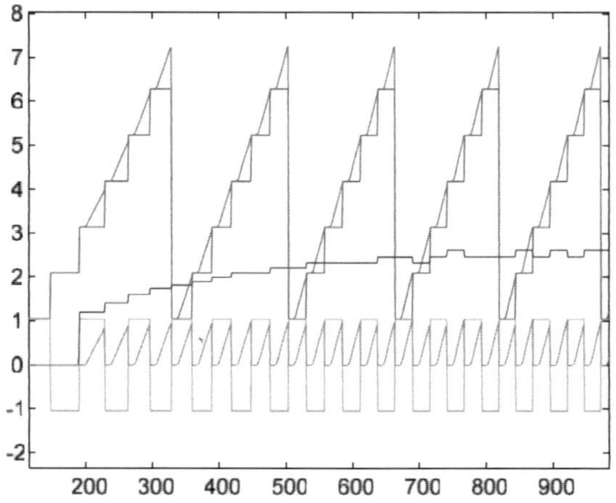

Figure 11. Stator current, Back EMF after interpolation

6. Conclusion

This paper represents a BLDC motor control system using interpolation derived from Hall sensors. The estimation technique is applied to a brushless DC motor with trapezoidal back EMF and shows the performance of the selection method for speed. The algorithm provides the implementation of PMSM vector control based on signals from discrete Hall detectors. With this, information about the speed of rotation will be obtained if the Hall sensors do not change. Due to the implementation of PMSM synchronous mode, the algorithm works on the fault of the three components. From the experimental result it can be said that the use of Hall detectors and interpolation will be more effective in a steady state. The proposed method provides better performance in the switching process for example the motor can be in a stationary mode for 3 seconds, while the back EMF is 20% lower.

Acknowledgement

The author would like to thank the Research and Development Sector at the Technical University of Sofia for the financial support.

References

[1] S. Ozturk, "Modelling, simulation and analysis of low-cost direct torque control of pmsm using hall-effect sensors", 2005

[2] M. Schroedl, "Sensorless control of permanent magnet synchronous motors," Elect. Mach. Power Syst., vol. 22, no. 2, pp. 173–185, Mar./Apr. 1994.

[3] J. Bu, L. Xu, T. Sebastian and B. Liu, "Near-zero speed performance enhancement of PM synchronous machines assisted by low cost hall effect sensors," in Proc. IEEE APEC'98, 1998, pp. 68-74.

[4] P. Pillay and R. Krishnan, "Modeling, Simulation, and Analysis of Permanent-Magnet drives, Part II: The Brushless DC Motor Drive," IEEE Trans. on Ind. App., Vol. 5, No.2, March/April 1989, pp. 274 –279

[5] T.M. Jahns, "Torque Production in Permanent-Magnet Synchronous Motor Drives with Rectangular Current Excitation," IEEE Trans. on Industry Applications, Vol. IA-20, No.4, July/August 1984, pp. 803-813.

[6]. V. Hristov and D. Slavov, "Stand For Research And Optimization Of DC Electric Drives With Digital Program Control," 2023 7th International Symposium on Innovative Approaches in Smart Technologies (ISAS), Istanbul, Turkiye, 2023, pp. 1-5, doi: 10.1109/ISAS60782.2023.10391716

IOP Conf. Series: Materials Science and Engineering **1317** (2024) 012005 doi:10.1088/1757-899X/1317/1/012005

Coupled Tanks Level Control Based on Volume Balance Method

Georgi Ruzhekov

Faculty of Automatics, Technical University of Sofia, Bulgaria

E-mail: g_ruzhekov@tu-sofia.bg

Abstract. The laboratory bench "Coupled tanks" represents a model of real technological plans where the levels and flow rates of liquids must be controlled. Such training benches are very common and there is a very large number of publications related to their control. All publications work with a non-linear model based on the Bernoulli equation for ideal fluids, which is non-linear – the square root of the fluid height is present. This equation or system of equations is linearized and attempts are made to control such systems by means of linear models. This article shows that this approach is not correct and proposes a completely different one based on the volume balance method.

1. Introduction

The simplest nonlinear model based of Bernoulli equation is [1,2,3,4] (Daniel Bernoulli, Hidrodynamica, 1738):

$$\frac{dh(t)}{dt} = -\frac{a}{A}\sqrt{2gh(t)} + u(t), \tag{1}$$

where:

h is fluid level in the tank

A - cross-section area of the tank

- outlet area of the outlet valve

g - gravitational constant

a

u - pump control

From (1) follows:

$$A\frac{dh(t)}{dt} = -a\sqrt{2gh(t)} + Au(t). \tag{2}$$

Equation (2) can be simplified to:

$$\frac{dV(t)}{dt} = -k\sqrt{h(t)} + Qp(t) = -Qout + Fp\big(u(t)\big), \tag{3}$$

where $V(t)$ is the instantaneous amount of fluid in the reservoir, $\frac{dV(t)}{dt}$ is the fluid flow rate and represents the difference between the pump flow rate $Qp(t)$ and the flow rate of the flowing fluid . Pump flow rate is a function of pump control $Fp\big(u(t)\big)$.

$Qout = k\sqrt{h(t)}$

The formula for the flow rate of the outflowing fluid $Qout = k\sqrt{h(t)}$ is only valid when the outflow is laminar and is uniformly distributed over the entire area a, which in practice cannot be achieved besides the fluid must not be viscous as friction losses in viscous liquids are significant and cannot be ignored. Fluid flow control from the pump is highly non-linear and usually has a dead zone. As it can be seen from (3) the control of the level in the tank is determined by two completely different processes, which are highly non-linear and attempts at linearization and combination will not lead to a significant practical result. This process is also non-stationary because the ambient temperature and the temperature of the fluid affect all elements of the system. Taking into account the above, it is clear that all attempts to linearize (3) and use linear control methods will not lead to significant practical effects, regardless of the claims of many authors that they have achieved satisfactory results.

The parameters described above can be seen in Fig. 1. The laboratory bench is shown in Fig. 2.

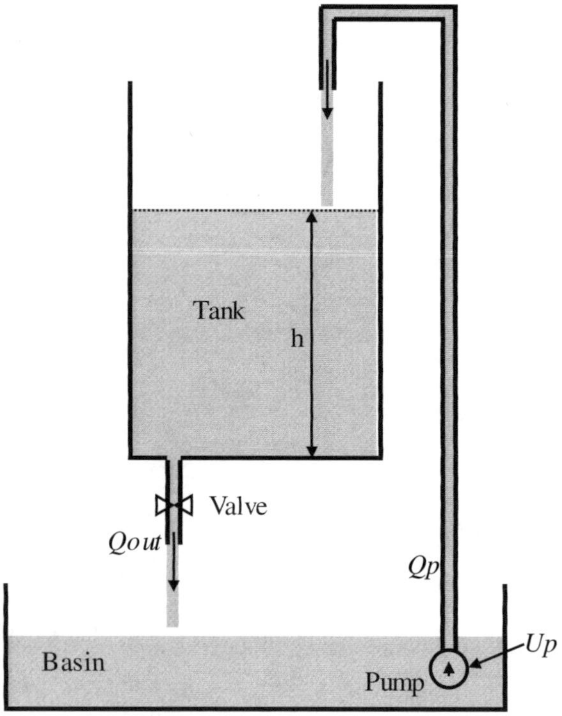

Figure 1. A simplified model of a tank

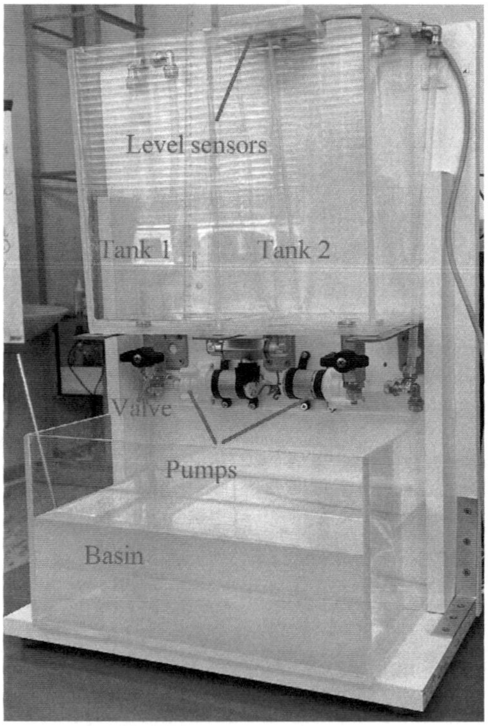

Figure 2. Laboratory bench

2. Volume balance method

The proposed method is based on the fact that in order to control the fluid level in the tank, it is necessary that the flow rate of the outgoing fluid is equal to the flow rate of the supplied fluid. For the implementation of this approach, it is necessary to initially implement the following steps to identify the parameters of the elements of the system:

Step 1. Determination of the flow rate of the pump depending on the control. For several fixed pump control values Up (e.g. 30, 50, 70, 100%), the pump flow is calculated according to (4):

$$Qp_j = A\frac{h_2 - h_1}{t_1 - t_2} = A\frac{\Delta h}{\Delta t}, \quad j = 1\cdots 4, Up_j = [30, 50, 70, 100], \%, \tag{4}$$

where Δh is the change in level with time Δt. These experiments are performed for fixed pump control values (30, 50, 70 and 100%). The pump cannot operate at less than 30% control, i.e. there is a dead zone. The levels h_1 and h_2 are set to 20 and 200 mm and the time to pass between the two levels is measured at the set pump control values. Calculated flow rates are saved in an array. The determination of pump flow rates is part of the developed control system and is started by pressing a button at the operator station. In these intervals, the pump flow rate is assumed to be relatively linear with respect to the pump control. The results of this experiment are shown in Fig.3.

Step 2. Determination of the output flow rate of the fluid depending on the level in the tank. 5 zones are defined by height. The tank is filled to the top and allowed to drain, while the average flow in each zone is being calculated (5). It needs to be taken into account that the debit is an outflow, i.e. it must not be omitted that it has a " - " sign.

$$Qout_i = A\frac{h_i - h_{i+1}}{t_i - t_{i+1}} = A\frac{\Delta h_i}{\Delta t_i}, \quad i = 1\cdots 5, \tag{5}$$

where:

$Qout_i$, $i = 1\cdots 5$ - fluid leakage rate for each interval

h_i are the set fluid levels in the tank.

Δh_i is the level change in the corresponding interval.

Δt_i - the time for this change.

Flow rates calculated in this way are recorded in an array, and their calculation in this way is part of the control system. Determination of the output flow rate is shown in Fig. 4.

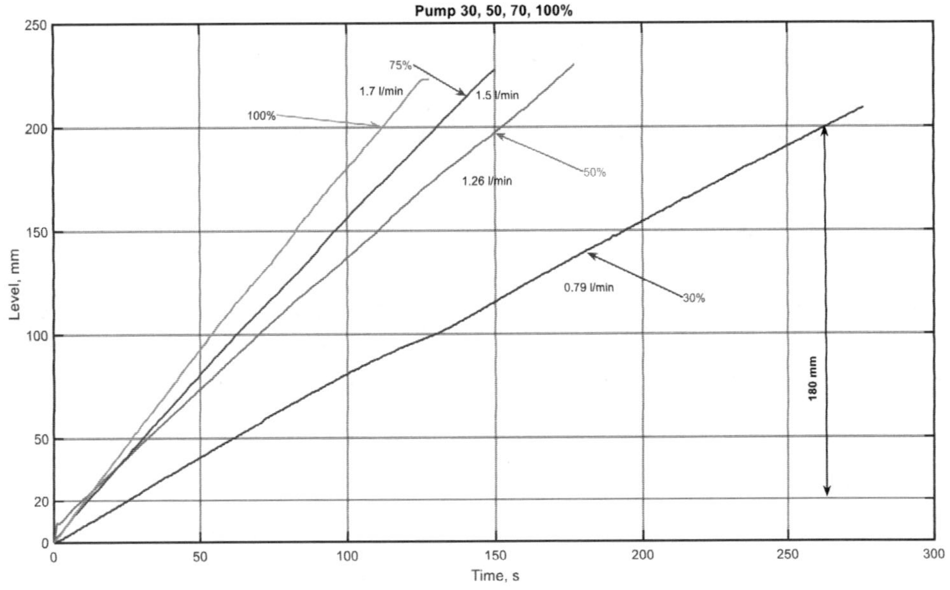

Figure 3. Determination of the flow rate of the pump

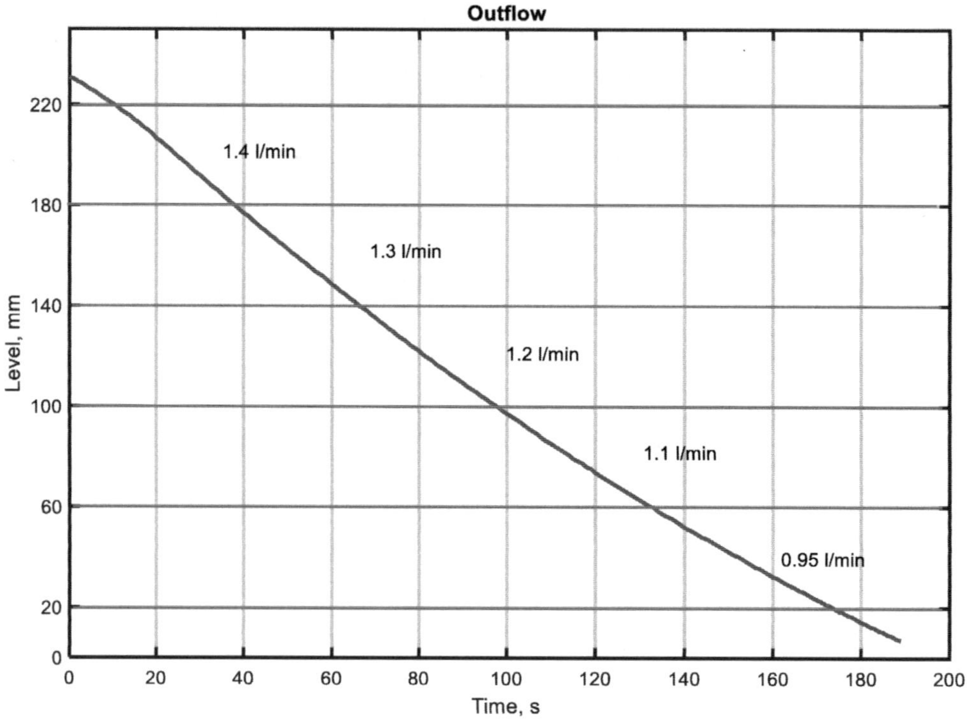

Figure 4. Determination of the output flow rate

3. Control algorithm

When using the volume balance method, the control algorithm varies depending on the difference between the setpoint (SP) and the current fluid level in the tank. Hysteresis is also introduced, $Hyst = |SP - PV|$ - an area between the setpoint and the current level. The hysteresis zone is chosen to be $1 \div 2\%$ of the tank height. The following cases are possible:

Case 1. The current level is lower than the setpoint minus the hysteresis $PV < SP - Hyst$ (below the hysteresis zone).
In this case, the pump control is set to the maximum value - $Up = 100\%$. This is the way to reach an area close to the setpoint (in the hysteresis area) as quickly as possible.

Case 2. The current level is higher than the setpoint plus the hysteresis (above the hysteresis area) $PV > SP + Hyst$.
In this case, the pump is turned off. This is the way to reach the hysteresis zone as quickly as possible. The arrival time depends on the outflow rate.

Case 3. In the hysteresis zone $PV - SP <= |Hyst|$.
The volume balance method requires that, in the hysteresis zone, the pump should deliver a flow Qp equal to the outflow $Qout$. Carried out experiments reveal what the output flow rate is at the relevant setpoint based on the pump experiments, the indicative control of the pump can be calculated for this flow rate.

IOP Conf. Series: Materials Science and Engineering **1317** (2024) 012005 doi:10.1088/1757-899X/1317/1/012005

The height interval in which the setpoint is located is determined from the table with the measured output flow rate (5). From there, the measured flow rate of the leaking fluid $Qout$ is taken.
For the value of leaking fluid $Qout$, the interval $j \div j+1$ is found for which $Qp_j <= Qout_i <= Qp_{j+1}$. In this interval, the pump control is determined by:

$$Up_i = \frac{Qout - Qp_i}{Qp_{i+1} - Qp_i}\left(Up_{i+1} - Up_i\right) + Up_i . \tag{6}$$

Thus, the flow rate of the pump will be close to, but not equal to, the flow rate of the flowing fluid. This is because the measurements of the flow rates (tank leakage and pump flow) were made at only a few intervals, as well as many other random factors. In this case, the difference between the flow rates is compensated by a PI(D) controller, the output of which is limited to a small range, e.g. $|Y_{PID}| < 15\%$. This output signal is added to the calculated pump control and thus corrects the pump flow rate:

$$Qp = Fp(Up_i + Y_{PID}) . \tag{7}$$

In this case, the PI(D) regulator is used for small corrections in the pump flow and works only in the hysteresis zone. The settings of the regulator are not critical to the operation of the entire system, they can be roughly calculated in advance and then fine-tuned.

4. Results

The experiments described above were carried out on a laboratory bench (Fig. 2), located in the Laboratory of Industrial Control Systems in the "System and Control Department", Faculty of Automation - TU-Sofia. A programmable logic controller Siemens CPU 1214C is used to control the processes. The pumps are PWM controlled. PWM should not be less than 30% (the pump cannot run at less power). This is a dead zone in the output of the regulator.
Pressure transmitters are used for level sensors. A software system was developed to calibrate the level sensors in two points – empty tank and full tank (230 mm). The achieved accuracy is about 1 mm, which is quite sufficient for this object.

On the basis of the experiments carried out to control the flow rates of the pump (4) and the outflow (5), a program system for level control was developed according to item 3 Control Algorithm (shown above). Hysteresis zone was set to 3 mm, setpoint was changed several times. The results of this experiment are shown in Fig. 5.
Analysis of experimental results:

- Transitions between levels are made as quickly as possible, because when the level increases, the pump works at 100%, and when it decreases, it is turned off.
- Set levels are maintained in the hysteresis zone (±3 mm), and the deviation is less than ±1.8mm.
- Control correction from the PID controller is within ±5%, indicating that the control algorithm is working correctly, that the flow rate measurement is correct, and that the controller is set correctly.

In Fig. 6 shows a trend from the operator station in working mode.

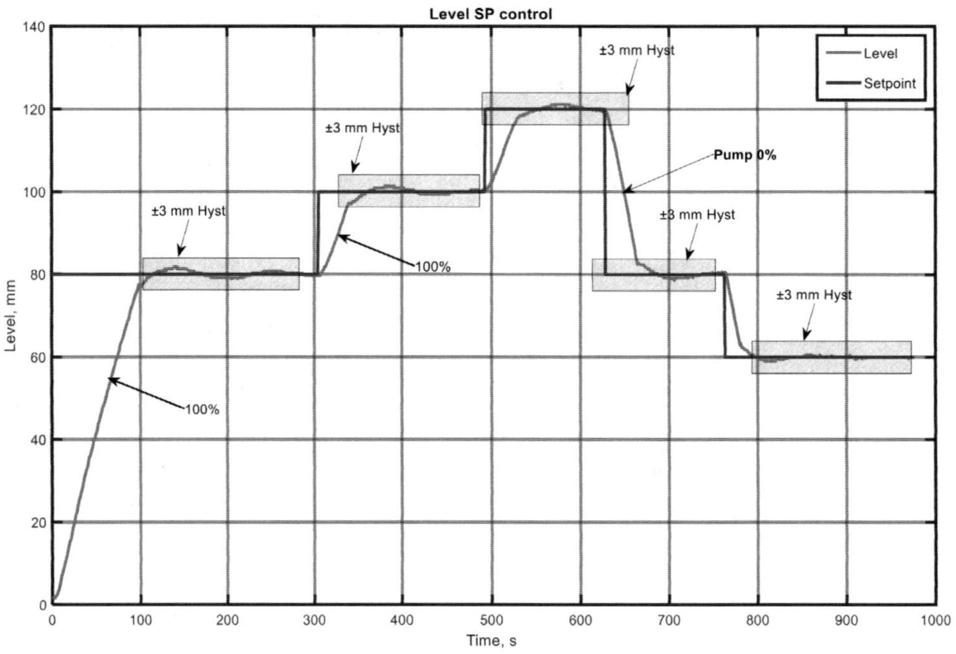

Figure 5. Real experiments

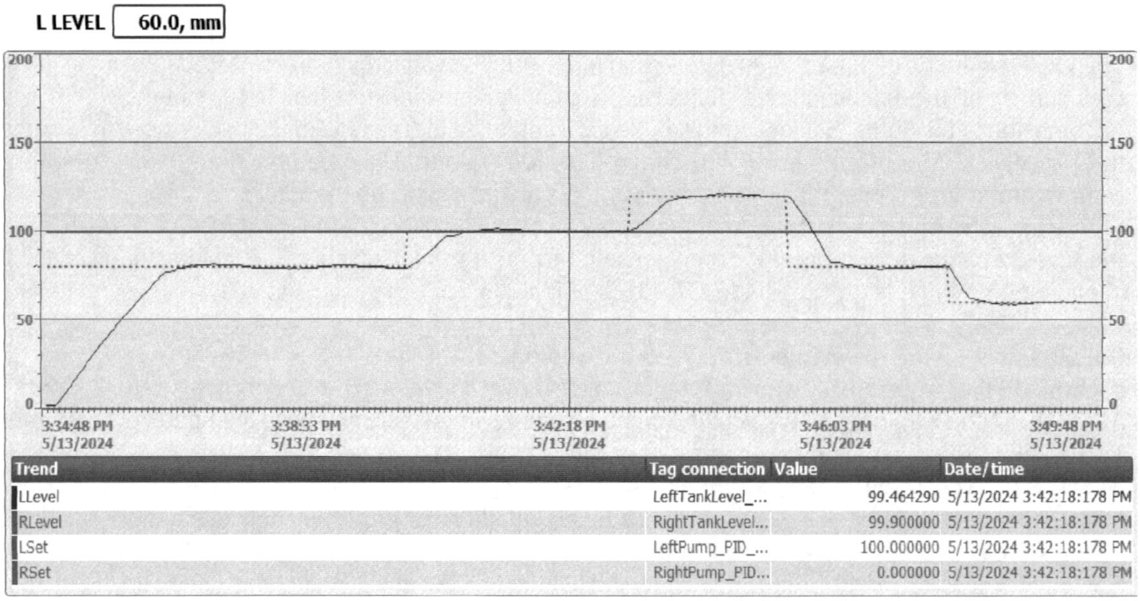

Figure 6. Trend from the operator station in working mode

Figure 7. Product tanks

This algorithm has been implemented in several industrial applications. The latter represents three product tanks with height of about 8 m and volume of about 25 m^3 are shown in fig. 7. They are part of a large industrial installation. There is a viscous product in them that is why temperature correction was also used. Powerful pumps are implemented and they are operated by inverters to control their debit. In all applications, the algorithm shown here works correctly.

5. Conclusion

A new approach to fluid level control in a tank is considered in the work. The paper reveals that methods considered in literature are not correct and cannot give adequate results, especially for higher tanks.
Necessary preliminary measurements are described to estimate delivery pump flow rates and tank fluid leakage. Control algorithm main steps are also presented.
The obtained results are shown to prove the system correct operation and the proposed method applicability. Other applications of this algorithm are also presented.
On the basis of this approach, algorithms can be developed to work with a change in the position (opening, closing) of the valve, which will also lead to a change in the flow rates (Fig. 1), to detect malfunctions in the system, etc. This approach can be also implemented in fluid flow rate control systems.

6. References

[1] Couple Tank Experiment, Quanser Inc., Markham, Ontario, 2012
[2] Level Control of Coupled Tanks System using Feedback Linearization Control Theory, Mohd. Tabrej Alam, Zohaib Hasan Khan, Integral University Lucknow, India, 2015
[3] Coupled Tanks, Control Experiments, Feedback Instruments Ltd., Park Road, Crowborough, East Sussex, 2010
[4] Coupled Tanks State Estimation Using a High Gain Like Observer, A.Adil, I. N'Doye, IFAC-PapersOnLine, Volume 54, Issue 14,2021, Pages 96-101

IOP Conf. Series: Materials Science and Engineering **1317** (2024) 012006 doi:10.1088/1757-899X/1317/1/012006

Forecasting of the Event-driven Processes Using LSTM Network in the Context of Time of Arrival of On-demand City Transport

Angel Marchev, Jr.[1*] and Boyan Lomev[2]

[1] Department of Management, University of National and World Economy, Sofia, Bulgaria
[2] Department of Statistics and Econometrics, "St. Kliment Ohridski" University of Sofia, Sofia, Bulgaria

[*]E-mail: angel.marchev@unwe.bg

Abstract. The precise forecasting of bus arrival times is an important element of implementing on demand city transport. This research uses of Long Short Term Memory (LSTM) networks for predicting bus arrival times in Sofia, Bulgaria. We evaluate the LSTM model against advanced models such as ARIMAX, VARX SARIMAX with Fourier terms Vector Autoregression, Bayesian Fourier models and Backpropagation Neural Networks using Root Mean Squared Error (RMSE) as the performance measure. The results points towards LSTM being better than approaches on routes by adeptly capturing intricate temporal relationships showcasing its promise in dynamic urban transit settings. Moreover the Bayesian Fourier model excels across routes underscoring its ability to accommodate trends effectively. These results suggest that incorporating LSTM networks into on demand transportation systems can notably enhance arrival time forecasts ultimately benefiting passengers and operational efficiency. This study demonstrates the usefulness of machine learning methods in urban transport planning and management by demonstrating how LSTM networks can improve optimization, on demand transportation.

1. Introduction

In order to enhance, on demand travel services it is important to integrate systems that will serve as the technological backbone of a reasonable control framework for a large citys transport system. The key components explained by Netov and Lomev (2022) include;

1. Prediction Model for Fixed Station Arrivals; This statistical model forecasts the travel time between fixed stations for the hour. It will utilize elements along with factors such as weather conditions, seasonal variations and videos from traffic cameras.

2. Simulation Model for Passenger Flow Demand. This model is predicts passenger behavior by developing probabilistic agent models, which will be used to simulate scenarios and identify demand patterns across locations and time slots.

3. Integration of Virtual Stations in the Model. Upon considering fixed stations and demand density potential virtual stations will be proposed within the model framework. The inclusion of these stations will adhere to constraints set forth by the Urban Mobility Center and Sofia Municipality.

4. Vehicle Routing Algorithm for Transport Optimization; This algorithm focuses on optimizing the routes taken by buses based on adjustments derived from input parameters.

5. Integration of Air Pollution Analysis in the Model. This aspect aims to assess relationships, between traffic patterns and air pollution levels through sophisticated urban simulations.

To expand the model to address air pollution it would require incorporating a representation of pollution, within traffic estimating car emissions and integrating a weather conditions model.

The current study primarily focuses on developing a model for predicting the arrival times at fixed stations. The goal is to outline the methodological considerations in predicting travel times between each pair of stops, utilizing an LSTM network. The additional objective involves conducting an comparative analysis with other models employed and suggesting the promising one.

Through a review of existing literature researchers have identified models for forecasting target data. Most of these models are elaborated in works by Marchev, Haralampiev et al. (2022a and 2022b) with models featured in the framework of this study. All models were evaluated using use case data.

- Autoregressive moving models with input (ARIMAX)
- Vector Autoregression with exogenous input (VARX)
- Seasonal autoregressive moving average models with exogenous input incorporating Fourier terms (SARIMAX w/ FT model) – noted as being methodologically superior, for forecasting cyclical time series data exhibiting various forms of seasonality (Marchev, 2023).
- When dealing with multivariate data in the study VAR models are expected to perform
- To account for seasonality using Fourier terms in modeling is deemed essential (Haralampiev, 2023).
- BpNN models serve as a methodology to utilize a non parametric model (Lomev, Marchev, 2023).
- LSTM neural networks, for long short term memory.

2. Event Driven Prediction

Event driven forecasting is a technique in analytics that focuses on predicting trends, behaviors or outcomes based on specific events or triggers. Unlike methods for analysis of time series, event driven forecasting focuses on distinct and temporally irregular events such as geopolitical shifts, natural disasters, product launches or market shocks.

The event driven forecasting offers an adaptive approach to foreseeing future outcomes in rapidly and irregularly evolving environments.

1. Event Recognition. Delineating events, with potential to affect predictions. Events can fall into two categories:

- Relevant events – they are under study to be forecasted;
- External events – events that impact the broader system or have global significance.

2. Gathering and Combining Data. Collecting data relevant to identified events which could include various modalities of data real time information sources. Integrating this data with existing datasets to improve forecasting models.

3. Modeling Events. Creating models that measure the characteristics of events. This involves using methods, machine learning models and scenario analysis to understand how an event influences the predicted variable.

4. Running Simulations and Analyzing Scenarios. Developing scenarios based on future events and assessing their consequences through simulations. This helps in grasping the range of outcomes and preparing for possibilities.

5. Monitoring in Real Time and Making Adjustments; Continuously keeping an eye out for developments and updating forecasts accordingly. This proactive approach ensures that forecasts stay current and accurate by incorporating information as it arises.

Common techniques utilized in event driven forecasting include:

- Machine Learning; Employing algorithms to recognize patterns and connections, between events and outcomes. Methods such, as networks, decision trees and support vector machines can be utilized.
- Natural Language Processing (NLP); Examining text information from sources like news, social media and reports to identify and measure occurrences and their emotional tone.
- Time Series Analysis with Exogenous Variables (ARIMAX, VARX); Expanding conventional time series models to factor in incidents as factors.
- Bayesian Models; Integrating knowledge. Adjusting predictions as fresh data emerges, making them fitting for dynamic and uncertain settings.
- Agent Based Modeling; Simulating the interactions of agents (such as consumers or companies) to grasp the impact of incidents on the entire system.

In the context of the current paper we make a distinction between time-series data and event-driven data in regard to bus arrival times. By definition time-series data is a set of discrete sequential and regularly time-spaced data-points. Whereas the bus arrival times data, albeit sequential, it does not conform with the requirement of temporal regularity.

3. Overview of LSTM method

Long Short Term Memory (LSTM) networks are a form of network (RNN) crafted for predicting future events from sequences of varying lengths. LSTMs stand out in predicting, categorizing and handling data. Unlike ARIMA models restricted to time series data, RNNs like LSTMs can manage kinds of sequential data.

Though LSTM technology was introduced in the 1990s its efficiency as a forecasting tool has recently risen due to advancements in its capabilities. LSTM network is a type of neural network used for their capability to process sequences of observations.

LSTM nincorporate feedback loops to overcome the short memory issue commonly seen in RNNs. Standard RNNs often face challenges, with retaining information over time due to their reliance on information feed resulting in problems like vanishing gradient caused by the repeated use of the recurrent weight matrix. In LSTM models an identity function replaces the weight matrix. It is controlled by a series of gates with individual recurrent loops.

A typical LSTM unit consists of a cell that stores values over a defined time frame. The input gate manages information flow into the cell while the output gate handles outgoing information flow, from the cell. The forget gate controls information that needs to be discarded.Actully, using these gates boosts LSTMs capacity to capture relationships effectively.

Complex systems often display linear effects due, to external influences. By including these factors as variables LSTMs can better. Forecast these relationships.

Once an output is calculated it is looped back into the model to predict the next value in the sequence. However each prediction introduces some level of error into the model. To address issues related to escalating gradients, we use sigmoid and tanh activation functions, so to regulate inputs and outputs for stabilizing values.

LSTMs demand more resources than other RNNs because there is an increased number of parameters used for the prediction process.

Recent studies have explored machine learning approaches to improve accuracy in predicting bus arrival times employing Long Short Term Memory (LSTM) networks.

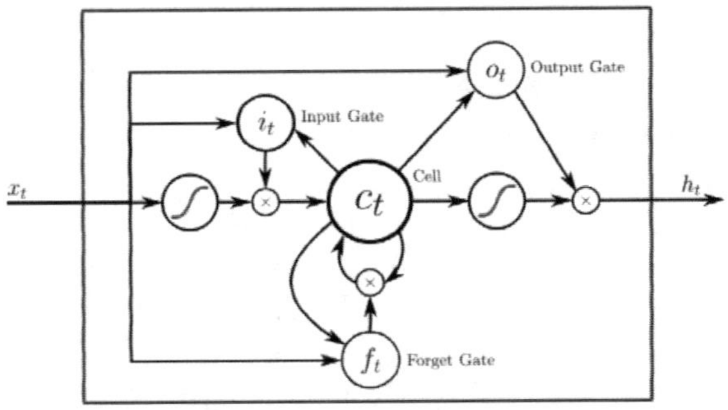

Figure 1. Long Short-Term Memory cell architecture (Graves, et al., 2013)

Petersen et al. (2019) introduce a system for forecasting bus travel durations that leverages spatio temporal correlations, within urban bus networks. They suggest using a neural network, with Convolutional and LSTM

IOP Conf. Series: Materials Science and Engineering **1317** (2024) 012006 doi:10.1088/1757-899X/1317/1/012006

layers to predict travel times over multiple outputs and time steps. This model's performance is compared to used methods for predicting bus arrival times.

In another research Zhou et al. (2019) a recurrent neural network is employed to forecast bus arrival times. The study demonstrates that incorporating factors like traffic conditions and road closures significantly improves prediction accuracy.

Moreover Seitbekova et al. (2020) outlines an approach that utilizes GPS data and real time road conditions to predict bus arrival times. The authors break down the prediction into time at bus stops and travel time between stations forecasting each part using LSTM neural networks.

Finally Li et al. (2020) introduce a model combining the strengths of LSTM and gradient boosting techniques to improve the precision of bus arrival time predictions.

These studies underscore the effectiveness of utilizing LSTM networks for predicting bus arrival times particularly when integrated with machine learning methods and dynamic variables.

4. Methodological Description of the LSTM Network Architecture

A. Data Preparation

1. Data Collection:
- Bus-stop arrival Data
- Traffic Data: Real-time traffic conditions.
- Weather Data: Weather conditions affecting travel.
- Schedule Data: Planned bus arrival times.
- Event Data: Incidents affecting travel (e.g., roadworks, accidents).

2. Feature Engineering: Construct features from raw data – see Marchev (2023), and Haralampiev (2023) for comprehensive list of derived features, including:
- datetime features (i.e. time of day, day of week, day of year, mounth of year, etc.);
- Fourier terms on daytime features;
- sequential bus course features (i.e. course of day, course of week, etc.);
- weather features (i.e. pour, clouds, etc.);
- dummy features (i.e. lockdowns, outliers, etc.)
- previous timings features (i.e. stay ot previous stop, delay, etc.)
- lagged-target features

3. Normalization: Normalize features to a range [0, 1]:

$$x' = \frac{x - min(x)}{max(x) - min(x)}$$

4. Sequence Creation: Create sequences of fixed length (L) for LSTM input:

$$X = X_{t-L+1}, X_{t-L+2}, \ldots, X_t$$

with the target variable (y_{t+k}) representing the arrival time at the next stop.

B. LSTM Network Architecture

The LSTM network consists of several layers designed to capture temporal dependencies in the data. The architecture can be summarized as follows:

1. Input Layer:
- Takes in the sequences of feature vectors $x_{t-L+1}, x_{t-L+2}, \ldots, x_t$.
- This is the initial layer where data enters the network.

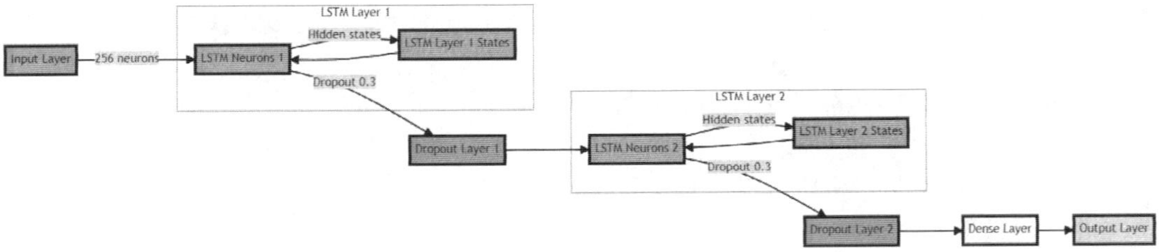

Figure 2. Principal architecture of the LSTM Network

2. LSTM Layer 1:
- Contains multiple LSTM neurons that process the input sequences.
- LSTM Layer 1 States:
- Each LSTM neuron maintains a hidden state $\left(h_t^{(1)}\right)$ and a cell state $\left(C_t^{(1)}\right)$ that are updated at each time step.
    - The hidden state captures the output at each time step, while the cell state captures the long-term dependencies.
    - The output of this layer is a sequence of hidden states that are passed to the next layer.

3. Dropout Layer 1:
- Applies dropout to the outputs of the first LSTM layer to prevent overfitting by randomly setting a fraction of input units to zero during training.

4. LSTM Layer 2:
- Contains another set of LSTM neurons that further process the sequences.
- LSTM Layer 2 States:
    - Each LSTM neuron in this layer maintains a hidden state $\left(h_t^{(2)}\right)$ and a cell state $\left(C_t^{(2)}\right)$.
    - These states are updated based on the outputs from the previous LSTM layer.
- This layer captures more complex temporal patterns and dependencies.

5. Dropout Layer 2:
- Applies dropout to the outputs of the second LSTM layer, similar to Dropout Layer 1, for regularization.

IOP Conf. Series: Materials Science and Engineering **1317** (2024) 012006 doi:10.1088/1757-899X/1317/1/012006

6. Dense Layer:

- A fully connected layer that maps the LSTM outputs to the target prediction space.
- Processes the final hidden states from the last LSTM layer to produce the output.

7. Output Layer:

- Produces the final prediction, such as the estimated bus arrival time $\widehat{y_{t+k}}$.

C. Architecture of the LSTM Neuron

1. Input x_t and Hidden State h_{t-1}: The LSTM neuron takes the current input (x_t) and the previous hidden state (h_{t-1}).

2. Concatenate: These inputs are concatenated to form a single vector ($[h_{t-1}, x_t]$).

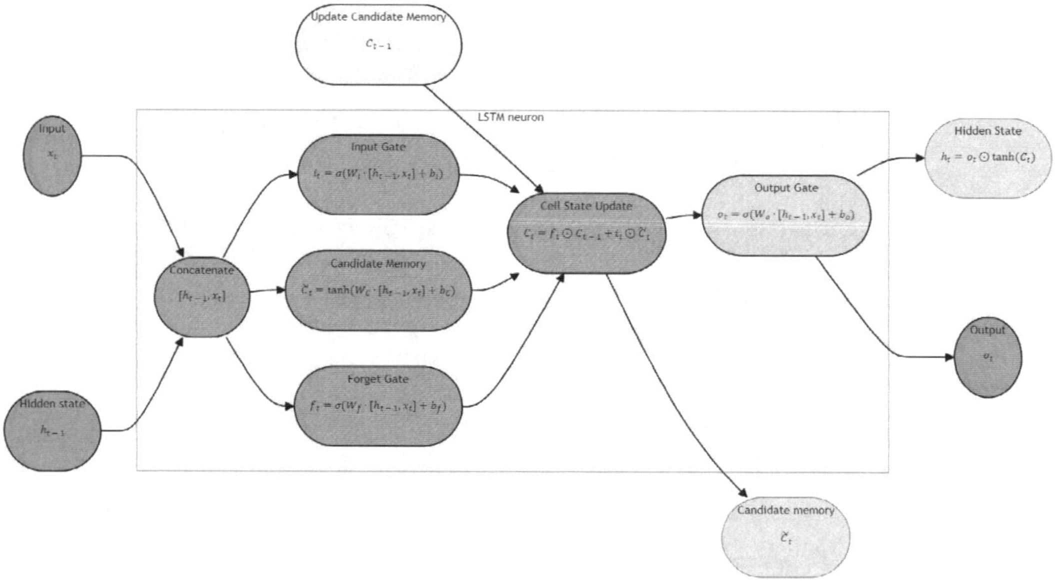

Figure 3. LSTM Neuron

3. Forget Gate: Computes the forget gate activation (f_t) which determines how much of the previous cell state (C_{t-1}) to retain.

$$f_t = \sigma\left(W_f \cdot \left[h_{t-1}, x_t\right] + b_f\right)$$

4. Input Gate: Computes the input gate activation (i_t) which controls how much of the new candidate memory ($\widetilde{C}_t$) should be added to the cell state.

$$i_t = \sigma\left(W_i \cdot \left[h_{t-1}, x_t\right] + b_i\right)$$

5. Candidate Memory: Computes the candidate memory cell state ($\widetilde{C}_t$).

$$\widetilde{C}_t = \tanh\left(W_C \cdot \left[h_{t-1}, x_t\right] + b_C\right)$$

6. Update Candidate Memory: This step explicitly shows that the candidate memory ($\widetilde{C}_t$) is updated by the input gate (i_t).

7. Cell State Update: Updates the cell state (C_t) using the forget gate, input gate, and updated candidate memory.

$$C_t = f_t \odot C_{t-1} + i_t \odot \widetilde{C}_t$$

8. Output Gate: Computes the output gate activation (o_t) to determine the next hidden state.

$$o_t = \sigma\left(W_o \cdot \left[h_{t-1}, x_t\right] + b_o\right)$$

9. Hidden State: Computes the new hidden state (h_t).

$$h_t = o_t \odot \tanh\left(C_t\right)$$

D. Training the Model

1. Loss Function: Mean Squared Error (MSE)

$$\text{MSE} = \frac{1}{n}\sum_{i=1}^{n}\left(y_i - \hat{y}_i\right)^2$$

2. Optimizer: Adam optimizer, which combines the advantages of Adaptive Gradient Algorithm (AdaGrad) and Root Mean Square Propagation (RMSProp).

E. Evaluation

Evaluate the model using Root Mean Squared Error (RMSE):

$$\text{RMSE} = \sqrt{\frac{1}{n}\sum_{i=1}^{n}\left(y_i - \hat{y}_i\right)^2}$$

By using LSTM networks, we can effectively model the temporal dependencies in bus arrival data and account for various influencing factors such as traffic and weather conditions. This approach provides a robust framework for predicting bus arrival times with high accuracy.

5. Empirical results

As an experimental setup we use a particular practical scenario. The input data for the models is sourced from two places - traffic data indicating arrival times at three stops along a bus route in Sofia (with additional information on schedules and stop durations) and weather data. The dataset includes times of arrival timings for four stops from January 10 2020 to July 29 2021. The observations, from the week (July 23 29 2021) are exclusively used as test data.

The target variable Y represents the duration of travel, between each stop. The variable is represented as a three column vector with 7394 data points for every instance of the vehicle passing through the course. To make the time series trend stationary first differences are computed. Some features are derrived such as time of day, day of the week, etc.

Each day there are, around 13 bus routes, which represent the unit of observation in the event series data. Some bus routes were not completed due to travel conditions or data recording errors making up 4% of the dataset. To ensure consistency in the data incomplete first and last days were excluded to focus on full day observations. The weather data is synchronized with the traffic data based on dates and times.

The optimal models for ARIMAX and VARX are chosen based on Akaike Information Criteria from a range of models. The Root Mean Square Error (RMSE) measured in seconds is used to evaluate and compare the forecast accuracy on the test set, which comprises observations from the test set. The software implementation is carried out using Matlab (R2020a).

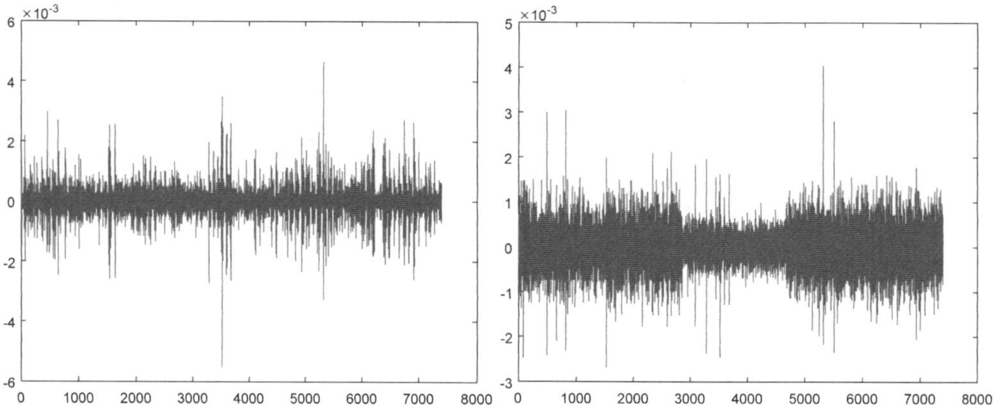

Figure 4. Stationary data for A) stop 2 (no mode shift in the data) and B) stop 1 (with mode shift in the data

A notable pattern in the data is evident from Figure 4. A behavior in arrival times at stop 2 during a period from August 1 2020, to December 31 2020.

Here are the results, from our experiments showing the Root Mean Square Error (RMSE) in seconds for each model at every stop:

- The Unconditional Mean acts as a reference point displaying performance across all stops but falling short compared to more intricate models.
- ARIMAX shows results to the mean with slight enhancements on certain routes but not consistently superior.
- VARX demonstrates progress over ARIMAX and unconditional mean in the two stops.
- LSTM Neural Network excels at Stop1 showcasing its ability to capture relationships effectively; it also performs well on other routes but not always topping the charts.
- SARIMAX with Fourier Terms emerges as the performer especially at Stops2 and3 indicating its limited effectiveness, in this scenario.
- Vector Autoregression (VAR) performs similarly to VARX with performance suggesting that incorporating exogenous variables (as, in VARX) offers some advantage.
- The Bayesian Fourier Model excels in performance for 2. Stop 3 showcasing its ability to capture trends effectively. It also delivers results across all routes nearly matching or surpassing the LSTM.

- The Backpropagation Neural Network demonstrates performance outperforming the mean but generally falling short compared to LSTM and Bayesian models.

Table 1. Experimental results – RMSE in seconds for every model for time of arrival at each stop

Models	Stop 1	Stop 2	Stop 3
Unconditional mean	36.17	21.25	47.78
ARIMAX	36.58	19.80	48.07
VARX	31.85	19.37	47.01
LSTM Neural Nets	29.94	18.61	48.06
SARIMAX w/ Fourier Terms	43.16	26.95	54.82
Vector Autoregression	32.03	19.45	47.43
Bayesian Fourier model	31.10	18.10	45.60
Back propagation NN	33.60	20.01	49.09

The Unconditional Mean serves as a baseline. It performs moderately across all stops but is outperformed by more complex models.

ARIMAX performs similarly to the unconditional mean, with slight improvements on some routes but not consistently better.

VARX shows an improvement over ARIMAX and unconditional mean, especially in the first two routes.

LSTM Neural Network has the best performance for Stop 1, indicating its strength in capturing complex temporal dependencies. It performs competitively for other routes, though not the best in every case.

SARIMAX w/ Fourier Terms is the worst performer overall, especially for Stop 2 and Stop 3. It seems less effective in this context.

Vector Autoregression (VAR) similar to VARX, but slightly worse in performance, indicating that including exogenous variables (as in VARX) provides some benefit.

Bayesian Fourier Model has the best performance for 2 and Stop 3, showing its strength in capturing seasonal patterns. It also performs very well across all routes, nearly as good as or better than the LSTM.

Backpropagation Neural Network performs moderately, better than the unconditional mean but generally worse than LSTM and Bayesian models.

6. Conclusions

The LSTM Neural Network excels in capturing patterns for routes (Stop 1) displaying potential for robust performance in various scenarios. Its strength lies in predicting bus arrival times due to its proficiency in handling data and learning long term dependencies.

The Bayesian Fourier Model boasts performance across multiple routes indicating a strong capability in capturing both short term fluctuations and seasonal patterns. It proves effective for routes with seasonal or periodic characteristics.

The primary recommendations suggest using LSTM Neural Networks or Bayesian Fourier Models, for bus arrival time predictions. These models exhibit notch performance and excel at managing complex temporal patterns.

Consider utilizing VARX if there is data influencing travel times. VARX and LSTM Networks undoubtedly outperform ARIMAX models.

Perhaps their strength lies in how well they can understand the connections, between pieces of information. It's best to steer of using SARIMAX with Fourier Terms for this job as it hasn't shown results, across all scenarios.

Acknowledgments

This publication is a result of our work on the project "Innovative demand responsive green public transportation for cleaner air in urban environment" (INNOAIR, INDEX: UIA05-202). This project is co-financed by the European Regional Development Fund through the Urban Innovative Actions Initiative.

The authors would like to present their gratitude to Faculty of economics and business administration at Sofia University for providing access to Matlab environment.

References

[1] Alex Graves, Abdel-Rahman Mohamed, and Geoffrey Hinton. (2013) Speech recognition with deep recurrent neural networks. In Acoustics, Speech and Signal Processing (ICASSP), 2013 IEEE International Conference on, pages 6645–6649. IEEE, 2013.

[2] Boyan Lomev, Angel Marchev; (2023) Application of artificial neural networks for predicting travel times for on-demand public transport in Sofia. AIP Conf. Proc. 11 December 2023; 2939 (1): 110001. https://doi.org/10.1063/5.0179198

[3] Angel Marchev; (2023) SARIMAX with Fourier terms for predicting travel times for on-demand public transport in Sofia. AIP Conf. Proc. 11 December 2023; 2939 (1): 110002. https://doi.org/10.1063/5.0179199

[4] A. Marchev, N. Netoff, K. Haralampiev, B. Lomev and K. Spassov, (2022a) "Innovative On-demand Public Transportation in Sofia - an Overview of the Algorithmic Tasks," 2022 10th International Scientific Conference on Computer Science (COMSCI), Sofia, Bulgaria, 2022, pp. 1-5, Doi: 10.1109/COMSCI55378.2022.9912578.

[5] Marchev Jr. Angel, Kaloyan Haralampiev, Boyan Lomev, (2022b) Predicting Travel Times for On-demand Public Transport in Sofia, IFAC-PapersOnLine, Volume 55, Issue 11, 2022, Pages 161-166, ISSN 2405-8963, https://doi.org/10.1016/j.ifacol.2022.08.066.

[6] Kaloyan Haralampiev; Application of Bayesian Fourier model for predicting travel times for on-demand public transport in Sofia. AIP Conf. Proc. 11 December 2023; 2939 (1): 110004. https://doi.org/10.1063/5.0178923

[7] Nikolay Netov, Boyan Lomev, Impact of digitalization of public transport and city mobility on traffic structure and on-road emissions, 6th European Conference on Electrical Engineering &

Computer Science (ELECS), Publisher: IEEE, 2022, pages:71-74, ISSN (print):978-1-6654-9780-0, ISSN (online):978-1-6654-9779-4, ISBN:978-1-6654-9779-4, doi:10.1109/ELECS55825.2022.00019, Ref, IEEE Xplore

[8] Petersen, N. C., Rodrigues, F., & Pereira, F. C. (2019). Multi-output bus travel time prediction with convolutional LSTM neural network, Expert Systems with Applications, Volume 120, 2019, Pages 426-435, ISSN 0957-4174, https://doi.org/10.1016/j.eswa.2018.11.028.

[9] Zhou, X., Dong, P., Xing, J., & Sun, P. (2019). Learning Dynamic Factors to Improve the Accuracy of Bus Arrival Time Prediction via a Recurrent Neural Network. Future Internet, 11(12), 247. doi: 10.3390/fi11120247

[10] Li, Y., Zhang, J., & Li, Z. (2020). Bus Arrival Time Prediction Using LSTM and Gradient Boosting. IEEE Transactions on Intelligent Transportation Systems, 21(5), 1231–1242. doi: 10.1109/TITS.2019.2961449

[11] Seitbekova, Y., Assilbekov, B., Beisembetov, I., & Kuljabekov, A. (2020). The bus arrival time prediction using LSTM neural network and location analysis. Journal of Advanced Transportation, 6(2), 46–57. doi: 10.1155/2020/6.2.1

IOP Conf. Series: Materials Science and Engineering **1317** (2024) 012007 doi:10.1088/1757-899X/1317/1/012007

Direct and inverse kinematics solutions of a particular type of 3-PSP parallel manipulator

Stanislav Enev[1*]

[1] Faculty of Automatics, Technical University of Sofia, Sofia, Bulgaria

*E-mail: enev@tu-sofia.bg

Abstract. In this paper, the direct and inverse kinematics problems for a particular type of 3-DOF, 3-PSP parallel manipulator are considered. The proposed direct kinematics solution is based on an existing one in the literature, but with an alternative, simplified second stage in the solution's formulation. An explicit analytical solution of the inverse kinematics problem is also proposed for one of the manipulator's operational modes, requiring one pass through the direct kinematics solution. The proposed solutions can be seamlessly extended to the inverted parallel structure.

1. Introduction

Parallel kinematic manipulators [1] design, modelling and control has been an active area of research for many years. PKM's advantages and drawbacks, within their application as manipulator structures, have long been identified, with research efforts aiming at tackling the associated modelling and control problems for given parallel structure or actively design new structures with inherent advantages within a pre-specified application [8]. 3-DOF PKMs, on their own, are being researched also within potential application as wrist implementing mechanism in larger "classical" series kinematics manipulator structures, as well as, as link module implementations for building highly modular hyper-redundant snake-like structures.

This paper considers a 3-DOF parallel manipulator with a particular type of a 3-PSP structure. Research on general 3-PSP manipulators can be found in [1], [5] and [6], focusing on kinematic problems' solutions, mobility and singularity analyses. The direct and inverse kinematics problems for the type of 3-PSP structure considered in this paper, are approached in [3], where two operational modes of the manipulator are identified. The inverse kinematics is formulated based on vector constraints written for the three legs of the mechanism, while the direct kinematics solution is obtained explicitly through geometrical analysis. In [4], the results are further developed, refined and augmented with velocity, singularity and workspace analyses. As noted in [4], the proposed inverse kinematics solution for the so-called there - "coupled, mixed-type" operational mode - which corresponds to operating the manipulator with two rotational and one translational degree of freedom of the moving platform, isn't suitable for real-time applications. Dynamic modelling and a particular approach to the control of the same manipulator can be found in [7].

In this paper the same mode of operation is investigated. Same approach is taken for obtaining the direct kinematics is proposed in [4], which consists of defining a sequence of homogeneous transformations to describe transitions between appropriately chosen coordinate frames. The end solution differs in its second part, where a different analytical geometry-based

IOP Conf. Series: Materials Science and Engineering **1317** (2024) 012007 doi:10.1088/1757-899X/1317/1/012007

approach is adopted to determine the coordinates of a point, to be the origin of the coordinate frame associated with the moving platform. In addition, a simplified analytical solution of the inverse kinematics is proposed, using one pass through the direct kinematics solution. The contributions of this paper can be summarized in the following:

- an analysis of the direct kinematics' structure is given, which leads explicitly to the definition of the two operational modes – the coupled mixed-type and the non-pure translational ones (operating the manipulator with three translational degrees of freedom);
- a line-intersection approach is followed in order to simplify the second part in the direct kinematics solution formulation - the way for getting to the origin of the moving plate's coordinate frame. This is achieved by taking the top plate's position as the Fermat point of the triangle formed by the 3-xSx centres' locations;
- a simplified explicit analytical solution for the inverse kinematics, for the coupled mixed-type operational mode, taking advantage of the direct kinematics' structure, is proposed.

Additionally, the inverse kinematics solution can be easily transformed for the inverted manipulator structure. The latter, in the coupled mixed-type operational mode, can be considered as an approximate 1-UP series kinematics mechanism, the manipulator, in the accepted as "regular" definition for base and moving ends, being seen as an approximate 1-PU series mechanism. The couple offers two alternatives for wrist mechanism implementation [9] in a larger manipulator structure, with substantially different workspace shapes that might suit the needs of different applications. It is also worth noting the possible application of the manipulator as a link mechanism in hyper-redundant manipulators, both enabled by the proposed inverse kinematics solution.

2. General description and considerations on the direct and inverse kinematics

The manipulator's structure is depicted in Fig. 1.

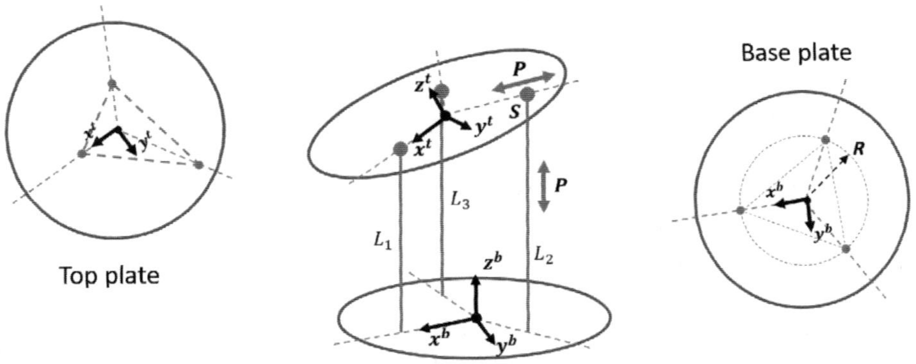

Figure 1. Structure of the considered 3-PSP parallel manipulator.

It consists of a base plate and top – moving plate connected through three identical "prismatic(P) – spherical(S) - prismatic(P)" joint series kinematic chains, forming a symmetrical structure. The prismatic (translational) joints associated with the base plate are perpendicular to it and are attached to the vertices of a unilateral triangle, the latter being inscribed in a circle with radius R. The circle's centre is chosen as the origin of the coordinate frame attached to the base, the z-axis chosen as the normal to the base plate plane, the x-axis pointing towards one of the point of

IOP Conf. Series: Materials Science and Engineering **1317** (2024) 012007 doi:10.1088/1757-899X/1317/1/012007

attachment of one of three chains to the base plate, labelled as "1". The prismatic joints associated with the moving plate lie in a plane (the plate's plane) intersecting at a single point and forming 120° angles. The intersection point is chosen as the position of the moving plate. A coordinate frame is attached to the moving plate with z-axis being normal to the top plate's plane and x-axis, aligned with the direction of the prismatic joint of chain "1". Without loss of generality, the distance between spherical and top plate's prismatic joint in each leg is taken as 0 (thus, the S-joints' centres lie in the x-y plane of the top plate's frame). It is easily shown that the intersection point of all three prismatic joints on the top plate is also the Fermat's point of the triangle formed by connecting the spherical joints' centres, a detail used in deriving the direct kinematics solution. Base plate's prismatic joints are actuated; their values, the manipulator's joint variables, being written in a vector for as follows:

$$L = [L_1 \quad L_2 \quad L_3]^{\boldsymbol{T}} \tag{1}$$

The following notations are used in the modelling related mathematical derivations:

- a superscript is used to indicate the coordinate frame in which a vector is expressed:
  $*^{\boldsymbol{t}}$ – for the coordinate frame attached to the top plate; $*^{\boldsymbol{b}}$ – for the coordinate frame attached to the base plate;
- $\boldsymbol{Rot}_{\boldsymbol{x(y,z)}}(\alpha)$ – homogeneous transformation matrix, describing a rotation abround the subscripted axis by an angle α;
- $\boldsymbol{Tr}_{\boldsymbol{x(y,z)}}(d)$ – homogeneous transformation matrix describing translation along the subscripted axis by a distance d;
- $\boldsymbol{H}_{\boldsymbol{A}}^{\boldsymbol{B}}$ – homogeneous tranformation matrix transforming frame $\boldsymbol{B}$ into frame $\boldsymbol{A}$; also, aligning frame $\boldsymbol{A}$ with frame $\boldsymbol{B}$.

New joint variables are introduced through the following transformation:

$$l = [l_1 \quad l_2 \quad l_3]^T, \tag{2}$$

with:

$$l_1 = L_1, \quad l_2 = L_2 - L_1, \quad l_3 = L_3 - L_2\,. \tag{3}$$

To get back in the original variables, we have:

$$L_1 = l_1, \quad L_2 = l_2 + l_1, \quad L_3 = l_3 + l_2 + l_1\,. \tag{4}$$

It is easily shown that this represents a valid, one-to-one bijective transformation. More importantly, it is seen that l_1 adds to all components of L the same value, so it only acts to produce translation in vertical direction of the top plate. Thus, changing l_1 should leave the rotation of the coordinate frame associated with the top plate unaffected, same for the x-, y- coordinates of the top plate's position. This, in turn, suggests the following structure of the homogenous transformation matrix for the forward kinematics in the new set of joint coordinates:

$$\boldsymbol{H}_{\boldsymbol{b}}^{\boldsymbol{t}}(l) = \boldsymbol{Tr}_{\boldsymbol{z}}(l_1).\begin{bmatrix} \boldsymbol{R}_{3x3}(l_2, l_3) & \begin{matrix} t_x(l_2, l_3) \\ t_y(l_2, l_3) \\ f_z(l_2, l_3) \end{matrix} \\ \boldsymbol{0} & 1 \end{bmatrix} = \begin{bmatrix} \boldsymbol{R}_{3x3}(l_2, l_3) & \begin{matrix} t_x(l_2, l_3) \\ t_y(l_2, l_3) \\ f_z(l_2, l_3) + l_1 \end{matrix} \\ \boldsymbol{0} & 1 \end{bmatrix}, \tag{5}$$

with:

- $\boldsymbol{R}_{3x3}(l_2, l_3)$ describing the orientation of the top plate's coordinate frame in the base plate's coordinate frame;
- $t_{x(y,z)}(l)$ – the subscripted coordinate of the top plate's position in the base plate's frame, with:

$$t_z(l) = f_z(l_2, l_3) + l_1. \quad (6)$$

Thus, two operational modes are possible, highlighted by the following bolded transformation matrix components:

$$\begin{bmatrix} \boldsymbol{R}_{3x3}(\boldsymbol{l}_2, \boldsymbol{l}_3) & \begin{matrix} t_x(l_2, l_3) \\ t_y(l_2, l_3) \\ \boldsymbol{t}_z(\boldsymbol{l}) \end{matrix} \\ \boldsymbol{0} & 1 \end{bmatrix} \text{ and } \begin{bmatrix} R_{3x3}(l_2, l_3) & \begin{matrix} \boldsymbol{t}_x(\boldsymbol{l}_2, \boldsymbol{l}_3) \\ \boldsymbol{t}_y(\boldsymbol{l}_2, \boldsymbol{l}_3) \\ \boldsymbol{t}_z(\boldsymbol{l}) \end{matrix} \\ \boldsymbol{0} & 1 \end{bmatrix}.$$

We can assign independently two rotational degrees of freedom and the top plate's position z-component (on the left) – it is checked in Section 4 that it is possible to assign the direction of the top plate's frame z-axis through the rotational degrees of freedom. This corresponds to the so-called in [4] "coupled, mixed-type" operational mode. The other possibility is to assign independently all three translational coordinates, that is the position of the top plate in the base plate's coordinate frame (on the right).

In the following, the first operational mode is considered.

3. Direct kinematics solution

The following figures are considered for formulating the direct kinematics solution.

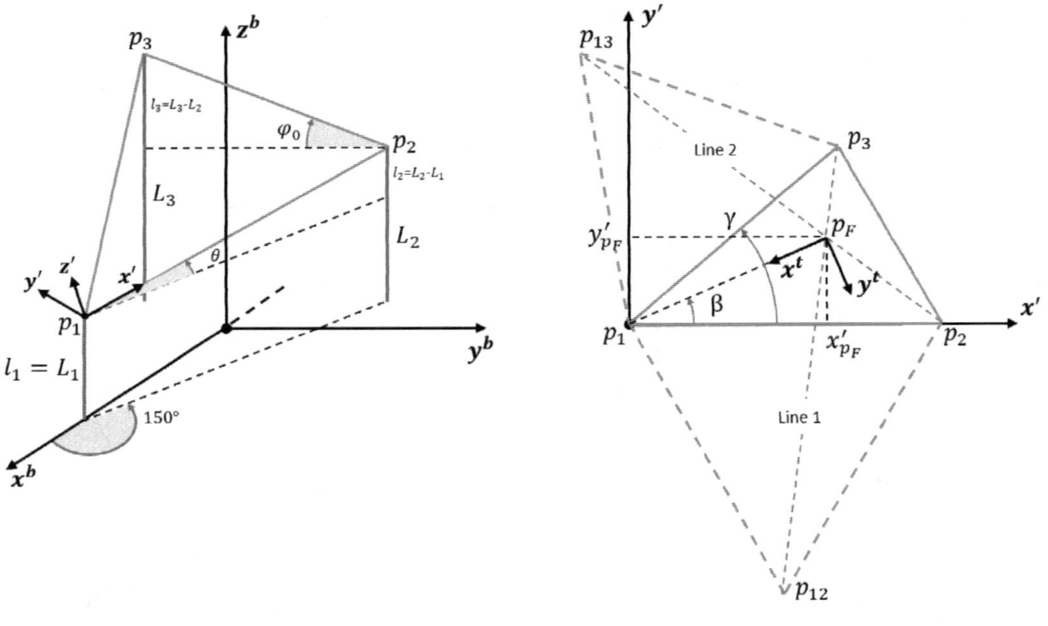

Figure 2. Coordinate frames used, angles and distances used for the transformations (on the left); Normal view of the top plate's plane.

An auxiliary coordinate frame - $\boldsymbol{f'}$ is defined as shown, the $\boldsymbol{x'} - \boldsymbol{y'}$ lying in the upper triangle's plane (being the top plate's plane and thus the $\boldsymbol{x^t} - \boldsymbol{y^t}$ plane).

The base frame $\boldsymbol{f^b}$ can be aligned with $\boldsymbol{f'}$ by the following sequence of elementary rotations/translations:

$$\boldsymbol{H}'_{\boldsymbol{b}}(l) = \boldsymbol{Tr_z}(l_1).\boldsymbol{Tr_x}(R).\boldsymbol{Rot_z}\left(\frac{5\pi}{6}\right).\boldsymbol{Rot_y}(-\theta).\boldsymbol{Rot_x}(\varphi), \tag{7}$$

with:

$$\theta = \text{atan}\left(\frac{l_2}{\sqrt{3}R}\right),\ \varphi_0 = \text{atan}\left(\frac{l_3}{\sqrt{3}R}\right),\ \varphi = \text{atan}\left(\frac{(l_2+2l_3).\cos(\theta)}{3R}\right). \tag{8}$$

The transformation aligning $\boldsymbol{f'}$ to $\boldsymbol{f^t}$ is given by:

$$\boldsymbol{H}^{\boldsymbol{t}}_{\boldsymbol{\prime}}(l) = \boldsymbol{Rot_z}(\beta).\boldsymbol{Tr_x}(|p'_F|)\boldsymbol{Rot_z}(\pi), \tag{9}$$

with: $\beta = \text{atan}\left({y'_{p_F}}/{x'_{p_F}}\right)$.

Thus, the transformation matrix from base frame to top plate frame can be written as:

$$\boldsymbol{H}^{\boldsymbol{t}}_{\boldsymbol{b}}(l) = \boldsymbol{H}'_{\boldsymbol{b}}(l)\boldsymbol{H}^{\boldsymbol{t}}_{\boldsymbol{\prime}}(l). \tag{10}$$

In order to determine the coordinates of p_F in $\boldsymbol{f'}$, p_3 coordinates are expressed in $\boldsymbol{f'}$ by:

$$\begin{bmatrix} p'_3 \\ \boldsymbol{1} \end{bmatrix} = \boldsymbol{H}^{\boldsymbol{b}}_{\boldsymbol{\prime}}(l)\begin{bmatrix} p^{\boldsymbol{b}}_3 \\ \boldsymbol{1} \end{bmatrix}, \tag{11}$$

with: $\boldsymbol{H}^{\boldsymbol{b}}_{\boldsymbol{\prime}}(l) = \boldsymbol{H}'_{\boldsymbol{b}}(l)^{\boldsymbol{T}}, p^{\boldsymbol{b}}_3 = \left[-\frac{R}{2},\quad -\frac{\sqrt{3}R}{2},\quad l_1 + l_2 + l_3\right]^T$; p_i – the i-th S-joint centre position.
We have the also:

$$p'_1 = [0,\quad 0,\quad 0]^T,\ p'_2 = \left[\sqrt{3R^2 + l_2^2},\quad 0,\quad 0\right]^T.$$

The auxiliary points positions are given by:

$$p'_{12} = \left[\frac{|p'_2|}{2},\quad -\frac{\sqrt{3}|p'_2|}{2},\quad 0\right]^T = \left[\frac{\sqrt{3R^2 + l_2^2}}{2},\quad -\frac{\sqrt{3}\sqrt{3R^2 + l_2^2}}{2},\quad 0\right]^T$$

and

$$p'_{13} = \begin{bmatrix} cos\gamma & -sin\gamma & 0 \\ sin\gamma & cos\gamma & 0 \\ 0 & 0 & 1 \end{bmatrix}\begin{bmatrix} |p'_3|/2 \\ \sqrt{3}|p'_3|/2 \\ 0 \end{bmatrix},$$

with $|p'_3| = \sqrt{\left(x'_{p_3}\right)^2 + \left(y'_{p_3}\right)^2}$, and $\gamma = \text{atan}\left({y'_{p_3}}/{x'_{p_3}}\right)$.

Lastly, p_F coordinates in $\boldsymbol{f'}$ are found as the intersection point of *Lines 1 and 2*, easily defined from the already determined point locations. Thus, the direct kinematics solution, in the form of (10) is completely specified.

4. Inverse kinematics solution

The inverse kinematices solution is formulated in two steps. First, a partial solution for the orientational component is found, thus determining l_2 and l_3 joint values. Then, the direct kinematics is calculated allowing to solve (6) for l_1, so that it is equal to the desired top plate's z-axis coordinate in the base frame.

The first part of the solution is based on the following observation:

$$\boldsymbol{H_b^t}(l) = \boldsymbol{Tr_z}(l_1).\boldsymbol{Tr_x}(R).\ \underbrace{\boldsymbol{Rot_z}\left(\tfrac{5\pi}{6}\right).\boldsymbol{Rot_y}(-\theta).\boldsymbol{Rot_x}(\varphi)}_{\boldsymbol{z-axis\ direction\ defining\ component}}\ .\boldsymbol{Rot_z}(\beta).\boldsymbol{Tr_x}(|p_F'|)\boldsymbol{Rot_z}(\pi),$$

The underlined section defining the orientation of $\boldsymbol{f'}$ in $\boldsymbol{f^b}$ (the z-axes of $\boldsymbol{f'}$ and $\boldsymbol{f^t}$ are pointing in the same direction by definition). For the corresponding rotation matrix, we get:

$$\boldsymbol{R_b'} = \boldsymbol{Rot_z}\left(\tfrac{5\pi}{6}\right).\boldsymbol{Rot_y}(-\theta).\boldsymbol{Rot_x}(\varphi) = \begin{bmatrix} -. & - & \frac{1}{2}sin\varphi + \frac{\sqrt{3}}{2}sin\theta cos\varphi \\ - & - & \frac{\sqrt{3}}{2}sin\varphi - \frac{1}{2}sin\theta cos\varphi \\ - & - & cos\theta cos\varphi \end{bmatrix} \tag{12}$$

the z-axis of $\boldsymbol{f'}$ in $\boldsymbol{f^b}$

Given the required top plate's z-axis orientation in the base frame – $(\boldsymbol{z'})^{\boldsymbol{b}}_{\boldsymbol{ref}}$ (it can be given within a rotation matrix generated through any of many known possible ways - Euler angles, Tait-Bryan angles sequences, quaternions, etc.)

$$\boldsymbol{R_{ref}} = \begin{bmatrix} r_{11} & r_{12} & r_{13} \\ r_{21} & r_{22} & r_{23} \\ r_{31} & r_{32} & r_{33} \end{bmatrix}, \quad (\boldsymbol{z'})^{\boldsymbol{b}}_{\boldsymbol{ref}} = [r_{13} \quad r_{23} \quad r_{33}]^T \tag{13}$$

solving for the third column (corresponding to the z-axis) is easy. For $sin\varphi$ and $tan\theta$ we have:

$$sin\varphi = \frac{r_{13} + \sqrt{3}r_{23}}{2}, \ tan\theta = \frac{\sqrt{3}r_{13} - r_{23}}{r_{33}}.$$

Given typical geometrical constraints of the mechanism, possible angle values lie in the range of -90° – 90°, so no "extraordinary" solutions are expected. Finally, from (8), for l_2 and l_3, we get:

$$l_2 = \sqrt{3}R.tan\theta$$
$$l_3 = \frac{1}{2}\left(3R.\frac{tan\varphi}{cos\theta} - l_2\right).$$

From (12) and (13), it is easy to find also:

$$\frac{tan\varphi}{cos\theta} = \frac{r_{13} + \sqrt{3}r_{23}}{2r_{33}},$$

which simplifies the calculations.

Finally, l_2 and l_3 values are plugged in (6) and l_1 is extraced, which completes the inverse kinematics solution.

5. Conclusion

In this paper, the direct and inverse kinematics problems for a particular type of 3-DOF, 3-PSP parallel manipulator are considered. The proposed direct kinematics solution is based on an existing one in the literature, but with an alternative, simplified second stage in the solution's formulation. An explicit analytical solution of the inverse kinematics problem is also proposed for an operational mode with two rotational and one translational degrees of freedom of the moving platform. The complete solution requires one pass through the direct kinematics solution. The proposed solutions can be seamlessly extended to the inverted parallel structure.

The possibilities for plugging the proposed inverse kinematics solution into a real-time control loop are to be further evaluated. Another direction for future research is related to the mobility characterization and workspace shape analysis for the inverted structure within an evaluation for its possible application as a wrist mechanism.

Acknowledgements

The author would like to thank the Research and Development Sector at the Technical University of Sofia for the financial support.

References

[1] Bonev IA., Geometric analysis of parallel mechanisms. PhD thesis, Laval University, 2002.
[2] Merlet P., Les Robots Parallèles. Hermès, Paris, 1997.
[3] Rezaei A, Akbarzadeh A, Mahmoodinia P, Akbarzadeh TM., Position, Jacobian and workspace analysis of a 3-PSP spatial parallel manipulator. Journal of Robotics and Computer-Integrated Manufacturing 29(4): 158–173, 2013.
[4] Mahmoodi Nia P., Akbarzadeh A., Close form solution for inverse and direct position analysis of a special 3-PSP parallel manipulator. In: 17th international conference on mechanical engineering, 2009.
[5] Di Gregorio R., Parenti-Castelli V., Position analysis in analytical form of the 3-PSP mechanism. Transactions of the ASME, Journal of Mechanical Design 2001;123:51–7.
[6] Gregorio, R.D., Parenti-Castelli V., Mobility Analysis of the 3-PSP Mechanism. In: Morecki, A., Bianchi, G., Rzymkowski, C. (eds) Romansy 13. International Centre for Mechanical Sciences, vol 422. Springer, Vienna, 2000. https://doi.org/10.1007/978-3-7091-2498-7_11.
[7] Toloue SF., Akbarzadeh M-R., Akbarzadeh A., et al., Position tracking of a 3-PSP parallel robot using dynamic growing interval type-2 fuzzy neural control. Applied Soft Computing 37: 1–14, 2015.
[8] Simoni R.; Carboni A.P.; Martins D., Enumeration of parallel manipulators. Robotica; Cambridge Vol. 27, Iss. 4, 589-597, Jul 2009.. DOI:10.1017/S026357470800497
[9] Shah D., Wu Y., Scalzo A., Metta G., Parmiggiani A., A comparison of robot wrist implementations for the iCub humanoid. Robotics 2019, 8(1), 11; https://doi.org/10.3390/robotics8010011

IOP Conf. Series: Materials Science and Engineering **1317** (2024) 012008 doi:10.1088/1757-899X/1317/1/012008

Electric field in a conductive medium – modelling and application

Ilona Iatcheva[1*], Denitsa Darzhanova[2] and Nikolina Petkova[1]

[1] Department of Theoretical Electrical Engineering, Technical University of Sofia, Sofia, Bulgaria
[2] Department of Electrical Measurements, Technical University of Sofia, Sofia, Bulgaria

*E-mail: iiach@tu-sofia.bg

Abstract. The paper presents some specific features of an electric field in a conductive medium, a theoretical base for numerical field modelling using the finite element method and its application in simulation of the processes in real electromagnetic systems. Application examples are presented, concerning simulations of the processes, taking part in contact resistance welding and insulation in the presence of conductive inclusions.

1. Introduction

A good and detailed knowledge of the processes, taking part in a given working system allows effective control and quality fulfilment of the requirements for the work process of this system. One of the most successful and modern methods of obtaining such knowledge is computer modelling of the processes in the studied system under different operating modes. In addition, the simulation of the device behaviour when changing the values of the design and operating parameters is an excellent opportunity for the possible optimization of existing devices and the synthesis of new, better ones.

The subject of consideration in this paper is the modelling of processes in systems operating on the basis of applying an electric field to given objects for the purpose of specific processing and impact on the structure of the parts of the object possessing conductive properties. As a result of this impact, currents flow in the conductive sections, the temperature state changes, changes occur in the properties of the materials and interaction forces arise, leading to qualitative changes in the microstructure of the treated objects.

The processing of various materials with the help of an electric field finds application in many fields [1-4]. Such type of treatment is applied to materials of both biological and non-biological origin. In some cases, these changes are necessary, desired and are the main purpose of the electric field treatment of the material, and in others some of these changes are a harmful side effect and should not exceed certain limits. The physicochemical changes in the state of materials treated with an electric field are the basis of many modern technologies. For example, the phenomenon of electroporation, which develops in this way, is widely used in the medicine, biology and food industry [5,6]. The processes taking part in creation of composite materials are also evolving in a similar way [7]. The principles used in high-tech technology of contact resistance welding lines are analogous [8]. The influence of the electric field acting on living nature often has harmful consequences. Also, a serious cause of aging and deterioration of insulation quality is application of a high voltage electric field to insulation containing moisture inclusions [9].

The paper presents two examples of numerical modelling of processes in systems and devices in which the treatment of materials with an electric field is carried out: contact resistance welding systems and modelling of the processes in electrical isolation, in the presence of conductive inclusions. The computer simulations of the processes in the considered systems are based on determination and analysis of the electric field distribution and the resulting changes in the thermal and mechanical state of the system, taking into account accompanying changes in the properties of the materials. Numerical modelling of the electric and thermal fields is performed using the finite element method (FEM).

2. Electric field in a conductive medium– specific features, equations, modelling possibilities

2.1 Electric field in a conductive medium– specific features

The electric field application on objects containing regions with conductive properties leads to the changes that in a number of cases can be very significant. The presence of electric field in areas containing conductive materials cause appearance of currents with density $\vec{J}$ whose magnitude and variation in time depend on the medium conductivity γ and on the value and time variation of the electric field strength $\vec{E}$ of the applied electric field. The flow of electric currents is associated with Joule losses Q, which are a sources of heat. The heat sources appearance leads to changes in the thermal state of the treated object, i.e. the electric field causes appearance of coupled thermal field. The change in temperature T is a factor influencing properties of the treated material and at temperatures exceeding certain limit values, additional changes in the structure of substances occur. These changes are even more pronounced when the object under the influence of an electric field is not homogeneous, but consists of areas of different material properties. Then at the boundary of the regions with different conductivity, a very sharp deformation in the electric field distribution is observed and significant mechanical stresses are present, which also affect the changing of the object's structure.

2.2 Electric field in a conductive medium - governing equations and mathematical formulation of the problem

As it was described above, the problem of electric field in a conductive medium is associated with a complex of coupled electric and possible thermal processes, combined with changes of the structure and physicochemical properties of the considered region. In the most general case, the mathematical formulation of the studied problems is based on the consideration of a time-dependent electric field, impact of which on the conductive materials leads to the appearance of a transient thermal field.

The dependencies (1) and equations (2) describe the processes of propagation of a time-dependent electric field in a region with dielectric permittivity ε and conductivity γ.

$$\vec{D} = \varepsilon \vec{E}$$
$$\vec{J} = \gamma \vec{E} \tag{1}$$

$$div\vec{D} = \rho$$
$$rot\,\vec{E} = 0 \qquad (2)$$
$$div\,\vec{J} + \frac{\partial \rho}{\partial t} = 0$$

Then taking into account that:

$$rot\,\vec{E} = 0 \quad \Rightarrow \vec{E} = -gradV \qquad (3)$$

In the equation (3), V denotes the scalar electric potential. Then after simple mathematical transformations, the equation that describes the distribution of the time-dependent electric field is obtained as:

$$div(\gamma\, gradV + \frac{\partial}{\partial t}(\varepsilon\, gradV)) = 0 \qquad (4)$$

In a case of time-harmonic electric field sinusoidally varying in time with frequency ω and when using complex quantities, imaginary unit j and ∇ operator, the equation takes the form:

$$\nabla.\left((\varepsilon - \frac{j\gamma}{\omega})\nabla \dot{V} \right) = 0 \qquad (5)$$

The following quantities are used in the description of the processes in the transient thermal field: T is temperature, k is thermal conductivity, ρ is density, C is heat capacity and Q is the heat source determined by the currents obtained in the electric field analysis.

Thus, the equation that describes the temperature field distribution in the studied area has the following form:

$$\rho C \frac{\partial T}{\partial t} + \nabla.(-k\nabla T) = Q \qquad (6)$$

2.3 Electric field in a conductive medium– modelling possibilities

The complex interactions and dependences between the interconnected coupled electric and thermal fields, propagating in a conductive medium, can be adequately modelled and analysed only if the changes in the structure and properties of the object during the process of treatment have been taken into account. A real possibility to fulfil these requirements is to model the fields distribution using FEM. It is important to consider separate consecutive time stages corresponding in time to the changes in the geometry, material properties of the object and the corresponding to the stage temperature.

The use of the finite element method for numerical modelling of the coupled electric and thermal fields is a suitable solution, since this method allows analysis of coupled field problems, as well as taking into account the nonlinear changes of the material properties. The presence of areas with complex geometry of the boundaries or change of the region's boundaries in the different stages of the analysis is not a problem, especially in 3D FEM modelling. In the examples presented below, the numerical modelling of the studied coupled electrical and thermal processes was carried out using COMSOL Multiphysics software package [10].

3. Examples

3.1 Modelling of the processes in a spot resistance welding system

The considered example presents the analysis of the coupled time-dependent electric and transient thermal field based on 3D field FEM modelling of the processes in a spot resistance welding system of reinforced steel bars [11].

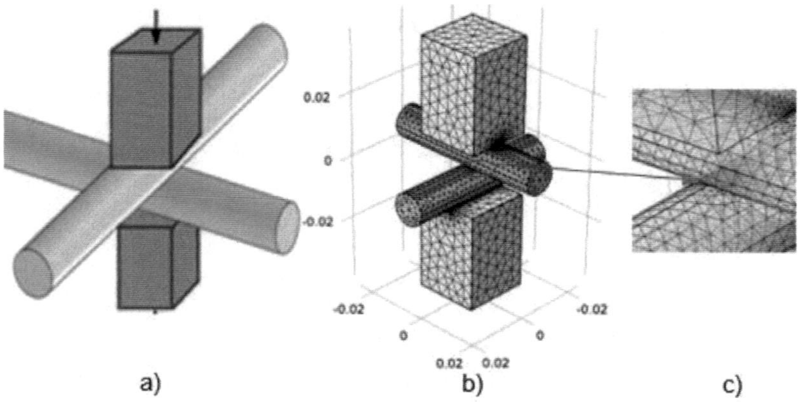

Figure 1: a) Spot resistance welding system; b) FE mesh discretisation; c) contact spot

The system consists of two electrodes and reinforced steel bars (Fig.1) that are to be welded. The two bars are held together under pressure exerted by the electrodes. The welding current is concentrated into a small contact "spot". Forcing a large current through the spot will melt the metal and form the weld. That permits the welding to occur without excessive heating of the remainder of the pieces. The system has been studied when 7kA AC current has been applied for 80 ms and contact area is heated by generated by resistance to flow of electric current. The work process causes significant changes. Because of the material softening, followed by melting and enlarging of the contact zone, the contact resistance changes significantly, as well as the applied power.

In order to account the changes of the contact zone size and contact resistance, three different stages of the welding process are considered in the study. They correspond to the dimensions of the steel bars contact area, contact resistance and time variation of the AC current: I stage: 0 – 0.003s, full power applied; II stage: 0.003 – 0.03s, half power applied; III stage 0.03 – 0.08s, quarter power applied. It should be noted that in the modeling, data obtained during experimental studies of the system were taken for verification.

Some results illustrating the numerical 3D FEM modelling of the time-dependent electric and transient thermal field are presented in Fig.2, Fig.3 and Fig.4. In Figure 2, the electric field is presented when current 7 kA is applied, the overlapping in the welded region reached 0.15mm and contact resistance is taken into account.

Figure 3 and Figure 4 correspond to the thermal state of the studied system during the treatment. As it can be seen the temperature in the contact region changes and at the moment *t= 80ms* the desired for the welding temperature of $2400^{\circ} C$ has been reached.

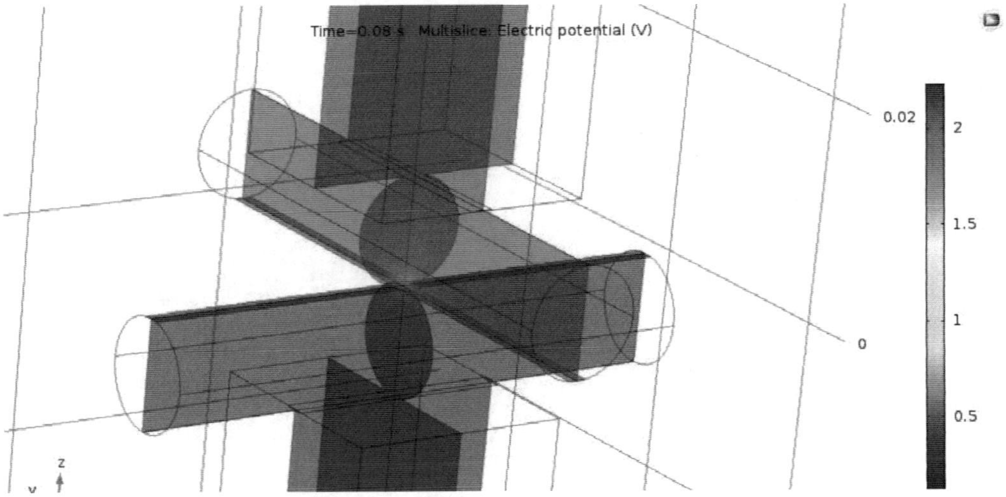

Figure 2. Electric potential distribution, 7 kA, 0.15mm overlapping

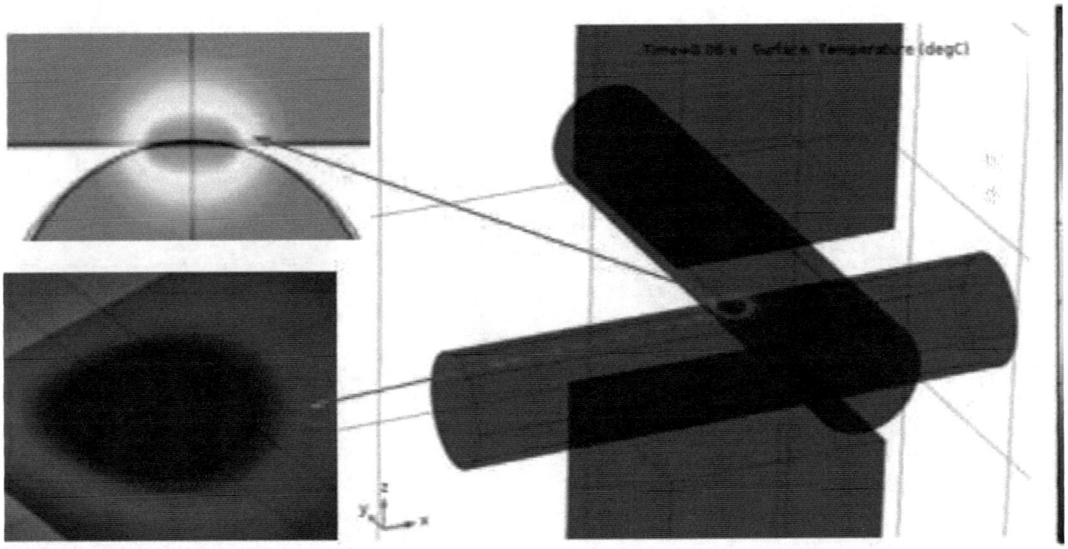

Figure 3. Temperature field distribution, 7 kA, 0.15mm overlapping

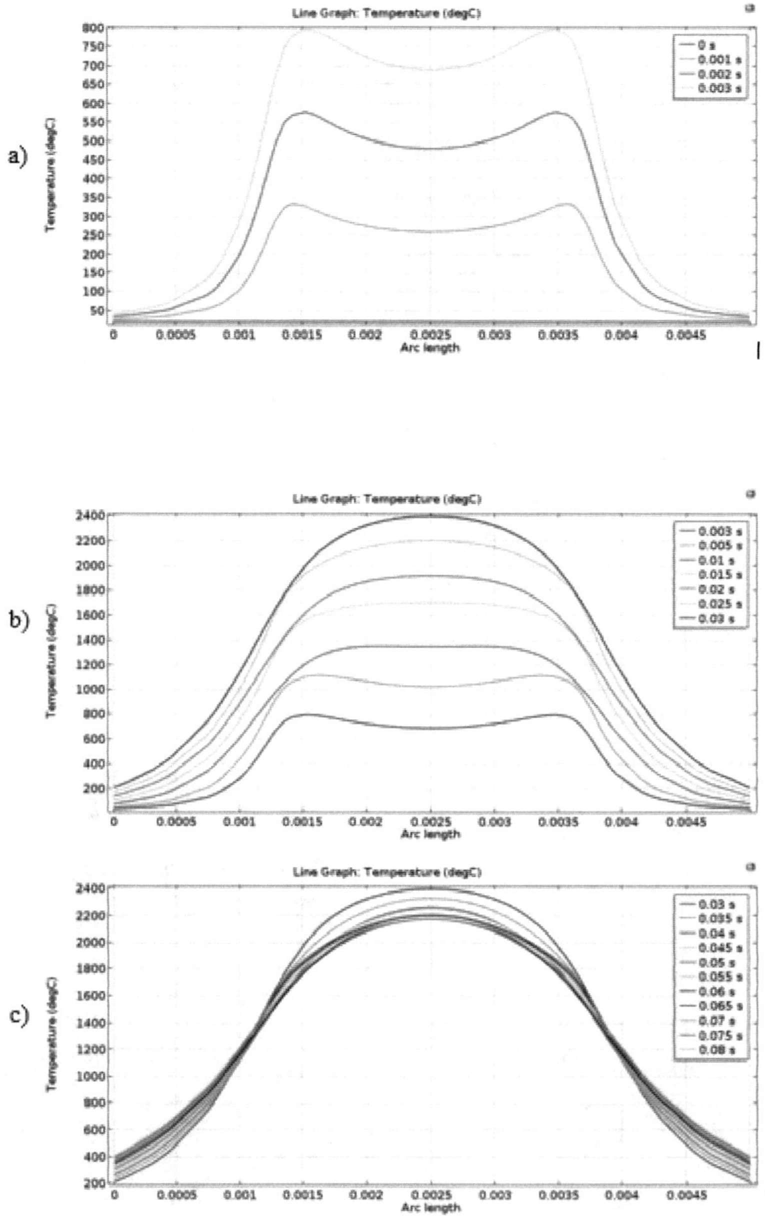

Figure 4. Temperature distribution in the contact spot area: a) I stage; b) II stage; c) III stage.

3.2 Modelling of the processes in electrical isolation

Elements of electrical insulation are subjected to various aggressive influences: strong electrical, thermal and mechanical loads. This causes aging and deterioration of insulation quality due to changes in structure, presence of various inclusions (conductive and non-conductive), partial discharges and change in breakdown dielectric strength, which determines the breakdown voltage.

In the present example, the distribution of the electric field in a simple electrode system, (Fig.5) is modelled and analysed.

IOP Conf. Series: Materials Science and Engineering **1317** (2024) 012008 doi:10.1088/1757-899X/1317/1/012008

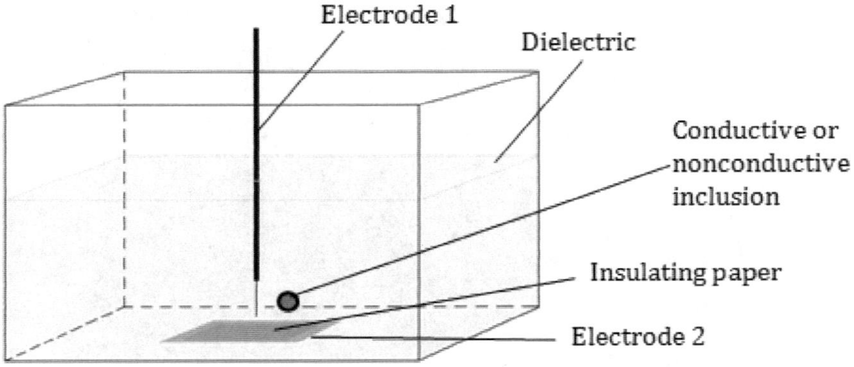

Figure 5. Studied system of electrodes in transformer oil.

The system is suitable for experimental investigations of insulation breakdown voltage [12]. It consists of two electrodes (needle and plate) placed in a transformer oil. It is assumed that the insulation is not uniform but contains single water inclusions in the form of spherical drops. This is one of the worst defects, which significantly affects the quality of the insulation and can cause a breakdown. The problem is that the presence of conductive inclusions significantly affects the electric field. Its distribution becomes highly non-uniform, the water drops line up along the electric field lines and form a conductive chain between the electrodes, in which a breakdown can occur.

The analysis of the electric field distribution is based on 3D field FEM modelling of the processes in studied system in case of insulation with no inclusion, insulation with only one water drop as an inclusion (Fig.6) and insulation with moisture inclusion consisting of several water drops (Fig.7).

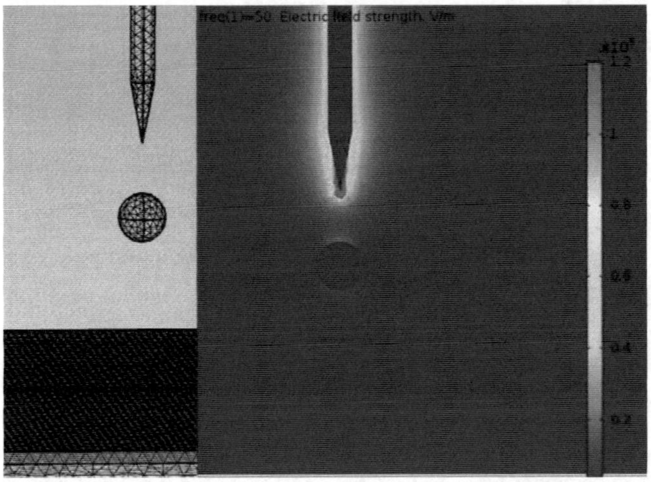

Figure 6. Presence of a single spherical water inclusion: part of the 3D mesh and electric field strength distribution.

IOP Conf. Series: Materials Science and Engineering **1317** (2024) 012008 doi:10.1088/1757-899X/1317/1/012008

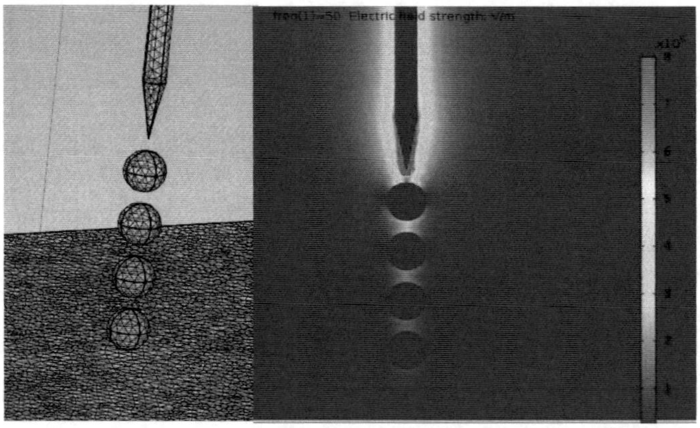

Figure 7. Presence of several water inclusions: part of the 3D mesh and electric field strength distribution.

As it can be seen from the figures illustrating the distribution of the electric field, the maximum value of the electric field intensity increases significantly with increasing water inclusions, exceeds the limit value corresponding to the breakdown intensity, and in this situation a breakdown of the insulation could occur.

4. Conclusions

The paper deals with some specific features and opportunities for numerical modelling of processes related to the application of a time-dependent electric field for the processing of material objects containing conductive zones. Examples are presented illustrating the proposed FEM modelling approach of the electric field distribution and, if necessary, the associated thermal field that appears due to the presence of the currents in the conductive parts.

References

[1] C. Jörgens and M. Clemens, "Modeling the Electro-Quasistatic Field of Ground Electrodes Under the Influence of Electro-Osmosis," 2020 IEEE 3rd International Conference on Dielectrics (ICD), Valencia, Spain, 2020, pp. 513-516, doi: 10.1109/ICD46958.2020.9341881.

[2] Böhmelt, S., Kielian, N., Hagel, M., Stiemer, M., Henkel, M.-L. and Clemens, M. (2020), "Electro-quasistatic field-simulation of biological cells using balanced domain-decomposition", COMPEL - The international journal for computation and mathematics in electrical and electronic engineering, Vol. 39 No. 3, pp. 739-755. https://doi.org/10.1108/COMPEL-10-2019-0414

[3] F. X. Hart and J. R. Palisano, 'The Application of Electric Fields in Biology and Medicine', Electric Field. InTech, May 23, 2018. doi: 10.5772/intechopen.71683.

[4] U. Musa, A. A. Mati, A. A. Mas'ud, G. S. Shehu, R. Albarracin-Sanchez and J. M. Rodriguez-Serna, "Modeling and Analysis of Electric Field Variation across Insulation System of a MV power Cable," 2021 International Conference on Electrical, Communication, and Computer Engineering (ICECCE), Kuala Lumpur, Malaysia, 2021, pp. 1-5, doi: 10.1109/ICECCE52056.2021.9514224.

[5] Novickij, V., Rembialkowska, N., Staigvila, G. et al. Effects of extracellular medium conductivity on cell response in the context of sub-microsecond range calcium electroporation. Sci Rep 10, 3718 (2020). https://doi.org/10.1038/s41598-020-60789-7

[6] S Dastani, K., Moghimi Zand, M., Kavand, H. et al. Effect of input voltage frequency on the distribution of electrical stresses on the cell surface based on single-cell dielectrophoresis analysis. Sci Rep 10, 68 (2020). https://doi.org/10.1038/s41598-019-56952-4

[7] T. Wong, I. Timoshkin, S. MacGregor, M. Wilson and M. Given, "Modeling of the Transient Electric Field in Multilayer Dielectric Composites Under Impulsive HV Energization," in IEEE Transactions on Dielectrics and Electrical Insulation, vol. 30, no. 1, pp. 220-229, Feb. 2023, doi: 10.1109/TDEI.2022.3214180
[8] Biele, L., Schaaf, P. & Schmid, F. Method for contact resistance determination of copper during fast temperature changes. J Mater Sci 56, 3827–3845 (2021). https://doi.org/10.1007/s10853-020-05490-w
[9] P. Pandey and C. C. Reddy, "Effect of Inhomogeneous Ageing of Insulation on Electric Field Distribution of a 220kV HVAC in-Service Cable," in IEEE Transactions on Dielectrics and Electrical Insulation, doi: 10.1109/TDEI.2024.3383412.
[10] COMSOL, Inc., COMSOL Multiphisics user's guide, version 5.3
[11] I. Iatcheva, D. Darzhanova, M. Manilova, Modeling of electric and heat processes in spot resistance welding of cross-wire steel bars. Open Physics, vol. 16, issue 1, pp. 1-8, 2018, doi: 10.1515/phys-2018-0001.
[12] I. Iatcheva and N. Petkova, "Study of electrical insulation in the presence of heterogeneous inclusions," 2016 19th International Symposium on Electrical Apparatus and Technologies (SIELA), Bourgas, Bulgaria, 2016, pp. 1-6, doi: 10.1109/SIELA.2016.7543012

IOP Conf. Series: Materials Science and Engineering **1317** (2024) 012009 doi:10.1088/1757-899X/1317/1/012009

Discrete Analysis of Electrical Circuit with Sources of Rectangular Pulses

Zhivko Georgiev[1], Ivan Trushev[2] and Atanas Chervenkov[3*]

[1] Fundamentals of Electrical Engineering Department, Technical University of Sofia, Bulgaria
[2] Fundamentals of Electrical Engineering Department, Technical University of Sofia, Bulgaria
[3] Fundamentals of Electrical Engineering Department, Technical University of Sofia, Bulgaria

E-mails: zhdgeorg@tu-sofia.bg[1],ivant@tu-sofia.bg[2], acher@tu-sofia.bg[3*]

Abstract. Discrete transients and stationary process are described in series RC-circuits when the supplied voltage contains periodic rectangular pulses. The analysis is performed using z - transformation to obtain analytical dependencies for the capacitive voltage in any time of the rectangular pulses. The approach is used for discrete analysis of various circuits in which rectangular pulses have different characteristics. The calculated capacitive voltages by the obtained discrete dependencies are identical, compared to the results obtained by SPICE simulation. The method proposed in the present work allows, together with the transient process, to investigate the corresponding stationary process. Cases of stable and unstable transients are shown.

1. Introduction

Some sources in different electrical and electronic circuits generate pulses with rectangular, triangular, saw-tooth forms. It is of interest from theoretical viewpoint and practical application in analysis of transients and conditionally stationary process in this circuit. The widely used Fourier series approach in cases of rectangular impulses it is not applicable to transient processes. This technique can only be used successfully for steady state processes for non-rectangular periodic voltages due to the Gibbs phenomenon [1, 2, 3]. Transient Analysis by Duhamel's Integral is possible for analogue signals and is difficult to apply to discrete analysis [4].

Method for analysis of the transients and conditionally stationary process, if power supply is periodic non-sinusoidal voltages, is applied here. How oi find voltage in capacitor, if power supply generates non-sinusoidal pulses is described in [5] using analytical dependences for the transients or steady-state. This discrete approach use z - transformation.

An input voltage *e(t)* constitute rectangular impulses. They are characterized by amplitude *E*, period *T* and time of impulset_i, respectively.

The purpose of the present study is to show the possibility of obtaining, through discrete analysis, strict analytical expressions, with the help of which the voltage on the capacitor can be calculated at any moment of the time of the transient process.

2. Base of the analysis

The base idea is to determine the discrete output capacitive voltage $u_C(t)$), of the usually electrical circuit (Fig.1) in discrete moment of time t = 0, t = T, t = $2T$, ..., t = nT, by taking samples from input voltage $e(t)$. For this purpose z-transformation is used to find transfer function, discrete impulse function and capacitive voltage [6, 7, 8, 9].

Analytically is possible to be imagined the power supply by Heaviside function

$$e(t)=\sum_{n=0}^{\infty} E\left[u(t-nT)-u(t-nT-t_i)\right], \quad (1)$$

and function $u(t)$ is the following

$$u(t)=\begin{cases}1, & t\geq 0\\ 0, & t<0\end{cases} \quad (2)$$

In this work first order series RC-circuit has been investigated.

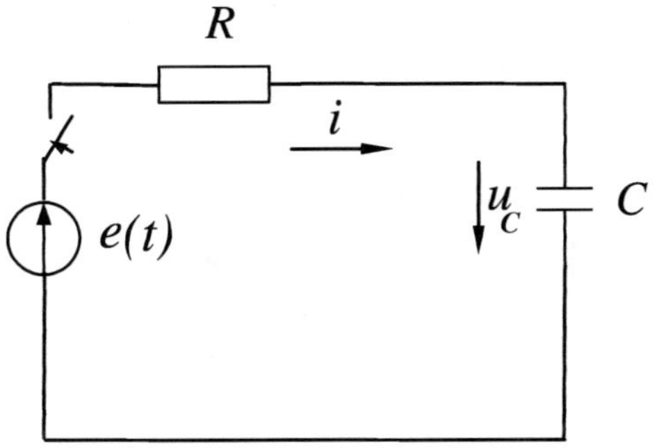

Fig. 1. An investigated circuit

The corresponding differential equation for the capacitive voltage is obtained [1, 2, 3]

$$RC\frac{\mathrm{d}u_C}{\mathrm{d}t}+u_C(t)=e(t) \ . \quad (3)$$

For the input voltage, for every period, the discrete input signal is

$$e[n]=Eu[n], \quad (4)$$

where $u[n]=1$ *for* $n=0, 1, 2, 3, \ldots$, $u[n]=0$ *for* $-1, -2, \ldots$

At the end z-image is implemented

$$\mathrm{E}(z)=E\frac{z}{z-1} \quad (5)$$

3. Discrete Analysis

The research is carried out according to the methodology described in [5].

A. *Determination of the impulse characteristic*

Let the supply voltage has the form $e(t)=E.u(t)$, , where E is constant. Let exist zero initial conditions, thenthe transient response is

$$h(t)=\frac{u_C(t)}{E}. \quad (6)$$

For the capacitive voltage expression is

$$u_C(t)=E\left(1-e^{-\frac{t}{\tau}}\right), \quad where\ \tau=RC\ . \quad (7)$$

IOP Conf. Series: Materials Science and Engineering **1317** (2024) 012009 doi:10.1088/1757-899X/1317/1/012009

For function of transfer we have

$$h(t)=\left(1-e^{-\frac{t}{\tau}}\right) \tag{8}$$

The impulse characteristic is [4,5]

$$h_i(t)=h(t)-h(t-t_i) \tag{9}$$

In the zero time is obtained respectively $h_i(0)=0$.

The next expression for the impulse characteristic $h_i(t)$ is obtained

$$h_i(t)=\begin{cases}\left(e^{\frac{t_i}{\tau}}-1\right)e^{-\frac{t}{\tau}}, & t>0\\ 0, & t=0\end{cases} \tag{10}$$

After discretisation impulse characteristic has the look

$$h_i[n]=\begin{cases}\left(e^{\frac{t_i}{\tau}}-1\right)e^{-\frac{nT}{\tau}}, & n=1,2,3,...\\ 0, & n=0\end{cases}. \tag{11}$$

B. Z – transformation of transfer function

Discrete impulse characteristic and its z – image is the transfer function:

$$H_i(z)=Z\{h_i[n]\} \tag{12}$$

The transfer function $H_i(z)$ is

$$H_i(z)=\sum_{n=0}^{\infty}h_i[n]z^{-n}=\sum_{n=1}^{\infty}h_i[n]z^{-n}=\left(e^{\frac{t_i}{\tau}}-1\right)\left(\frac{z-z+e^{-\frac{T}{\tau}}}{z-e^{-\frac{T}{\tau}}}\right)=$$
$$=\left(e^{\frac{t_i}{\tau}}-1\right)e^{-\frac{T}{\tau}}\left(\frac{1}{z-e^{-\frac{T}{\tau}}}\right) \tag{13}$$

C. An image of the capacitive voltage

The image of the output signal -capacitive voltage is

$$u_C(z)=E\left(e^{\frac{t_i}{\tau}}-1\right)e^{-\frac{T}{\tau}}\frac{z}{(z-1)\left(z-e^{-\frac{T}{\tau}}\right)}, \tag{14}$$

D. The discrete capacitive voltage

After decomposition in elementary fractions the capacitive voltage $u_C[n]$ is

$$u_C(z)=\frac{\alpha z}{(z-1)\left(z-e^{-\frac{T}{\tau}}\right)}=\frac{\beta_1 z}{(z-1)}+\frac{\beta_2 z}{\left(z-e^{-\frac{T}{\tau}}\right)} \tag{15}$$

The coefficients c_1 and c_2 by the undetermined coefficients method are determined.

$$\beta_1=\frac{\alpha}{1-e^{-\frac{T}{\tau}}},\qquad \beta_2=\frac{\alpha}{e^{-\frac{T}{\tau}}-1} \tag{16}$$

After that, the following equation for $u_C(z)$ is obtained

$$u_C(z) = \frac{\alpha}{\left(1-e^{-\frac{T}{\tau}}\right)}\left[\frac{z}{(z-1)} - \frac{z}{\left(z-e^{-\frac{T}{\tau}}\right)}\right]. \tag{17}$$

Finally the inverse Z transformations is applied

$$Z^{-1}\left\{\frac{z}{(z-1)}\right\} = u[n] = 1 \quad , n \geq 0, \tag{18}$$

$$Z^{-1}\left\{\frac{z}{\left(z-e^{-\frac{T}{\tau}}\right)}\right\} = e^{-\frac{nT}{\tau}}, \tag{19}$$

and for $u_C[n]$ is obtained

$$u_C[n] = \frac{E\left(e^{\frac{t_i}{\tau}}-1\right)e^{-\frac{T}{\tau}}}{\left(1-e^{-\frac{T}{\tau}}\right)}\left(1-e^{-\frac{nT}{\tau}}\right), \quad n = 0,1,2,... \tag{20}$$

After some transformations are obtained the next ones expressions for capacitive voltage

- For the time $0 \leq t' \leq t_i$ and a random number n of period

$$u_C(t',n) = E + E\left\{\frac{1-e^{\frac{t_i}{\tau}}}{1-e^{-\frac{T}{\tau}}}e^{-\frac{nT}{\tau}} - \frac{1-e^{-\frac{(T-t_i)}{\tau}}}{1-e^{-\frac{T}{\tau}}}\right\}e^{-\frac{t'}{\tau}}; \tag{21}$$

- For the time $t_i \leq t' \leq T$ and a random number n of period

$$u_C(t',n) = E\left\{\frac{1-e^{\frac{t_i}{\tau}}}{1-e^{-\frac{T}{\tau}}}e^{-\frac{nT}{\tau}} - \frac{1-e^{\frac{t_i}{\tau}}}{1-e^{-\frac{T}{\tau}}}\right\}e^{-\frac{t'}{\tau}}. \tag{22}$$

With these dependencies, the voltage can be determined anytime in transients and conditionally stationary process. Also the initial time *t* also can be obtained.

4. Applications of the discrete analysis

As the applications of the theory described above, the next cases are considered.

Case 1

Usually, in digital technology, rectangular signals are symmetrical with a pulse time of half the period T. Therefore, this is the most common case.

The parameters of the consider circuit are voltage level E=200mV, period T = 10ms, time of the impulse t_i = 5ms, resistance of the resistor R = 1kΩ, capacitance C = 10µF. Will be calculated the capacitive voltage at time t = 46ms using analytical formulas. Will be analyzed the circuit by PSPICE simulation for the capacitive voltage $u_c(t)$. Will be compared analytical results with simulation results.

Analytical Solution 1

First next values are calculated

$\tau = RC = 1.10^3.10.10^{-6} = 0.010,\quad T/\tau = 1.0,\quad 5T/\tau = 5.0,\quad (T - t_i)/\tau = 0.2,\ t_i/\tau = 0.5,$
$t'/\tau = 0.6.$

After that these data are substituted in expression (22) and the capacitive voltage is obtained

$$u_C(0.006,5) = 0.2\left\{\frac{1-e^{0.5}}{1-e^{-1.0}}e^{-5.0} - \frac{1-e^{0.5}}{1-e^{-1.0}}\right\}e^{-0.5} = 0.102V$$

Simulations 1

The capacitive voltage $u_C(t)$ is simulated by PSPICE and obtained graph is shown on Fig.2. The capacitive voltage at the time t = 46*ms*, computed by simulation is $u_C(46ms)$ = 102.5*mV*. The time t = 46*ms* refers to 5 period(n = 5) of the impulses and the corresponding time is $t' = 46-(5-1)10 = 6ms$. A time at which the transient is almost complete is used, which is usually after 4-5 time constants τ.

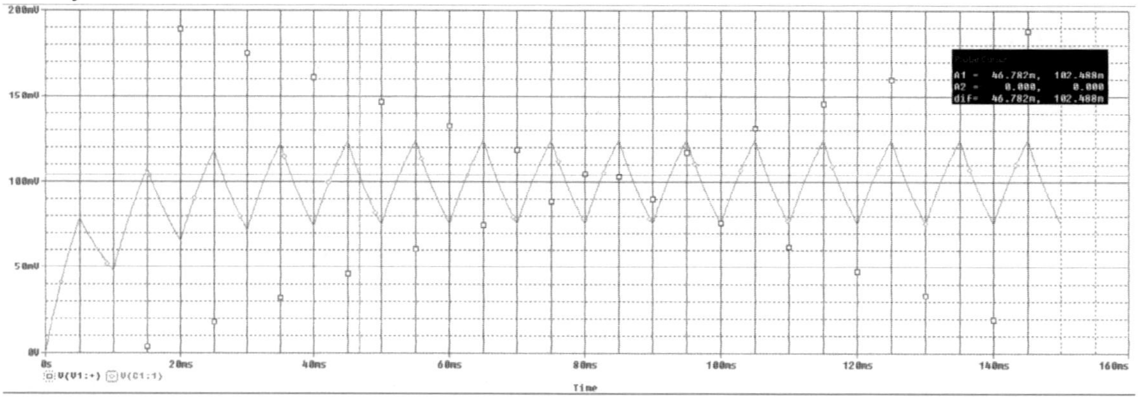

Fig. 2. A capaciive voltage obtained by simulations at $t_i = 0.5ms$

It is good match between the analytical results and the results, obtained by simulation.

On the Fig. 2 are shown both processes for capacitive voltage obtained by simulations - initial transient and the next steady-state process on the output voltage. It can be easily estimated that the analytical formula for the conditionally stationary process is such as mix of exponential functions.

Case 2

Pulse-width modulation is a widely used technique in controlling analog devices on digital outputs of microcontrollers. PWM is applied in many electronic communication devices, power control devices and in many of the converters [10, 11, 12, 13]. In PWM modulation, the pulse time t_i varies.

Analytical Solution 2

The parameters of the consider circuit are voltage level E=200mV, period T = 10ms, time of the impulse t_i = 1ms, resistance R = 1kΩ, capacitance C = 10μF. Will be calculated the capacitive voltage at time t = 46ms using analytical formulas. Will be analyzed the circuit by PSPICE simulation for the capacitive voltage $u_c(t)$. Will be compared analytical results with simulation results.

First next values are calculated

$\tau = RC = 2.10^3.13.10^{-6} = 0.010,\quad T/\tau = 1.0,\quad 5T/\tau = 5.0,\quad t_i/\tau = 0.1.$

After that, these data are substituted in expression (22) and the capacitive voltage is obtained

IOP Conf. Series: Materials Science and Engineering **1317** (2024) 012009 doi:10.1088/1757-899X/1317/1/012009

$$u_C(0.006,5) = 0.2\left\{\frac{1-e^{0.6}}{1-e^{-1.0}}e^{-5.0} - \frac{1-e^{0.6}}{1-e^{-1.0}}\right\}e^{-0.7} = 0.0159V$$

Simulations 2

The capacitive voltage $u_C(t)$ is simulated by PSPICE and obtained graph is shown on Fig.3. The capacitive voltage at the time t = 46*ms*, computed by simulation is $u_C(46ms)$ = 16.6*mV*. The time t = 46*ms* refers to the 5 period(n = 5) of the impulses and the corresponding time is $t' = 6ms$.

As well, simulation results and analytical solution has good match.

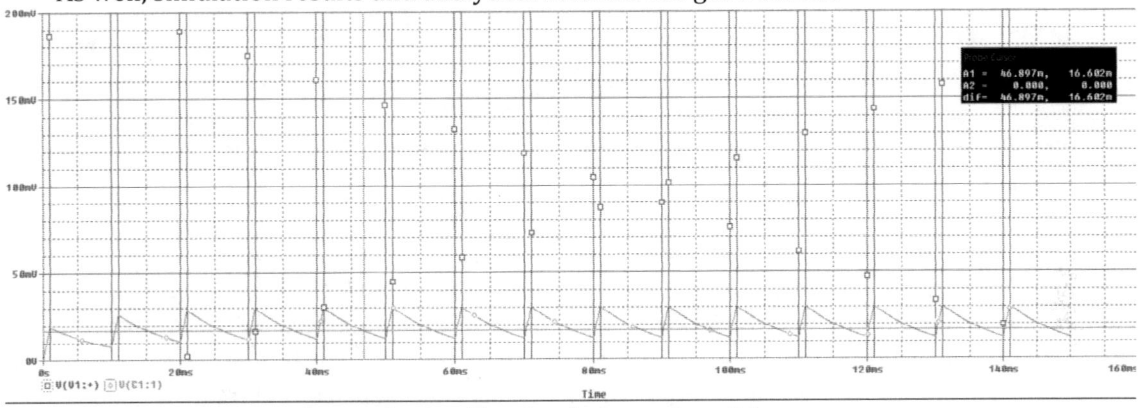

Fig. 3. A capacitive voltage obtained by simulation at t_i = 0.1ms

Unfortunately, a transient process does not develop in the circuit, and the steady-state process is unstable, because the voltage level of the capacitor is very low and cannot be used to control the power semiconductor elements in the rectifiers, signal amplifiers and in the converters.

Case 3:

The parameters of the consider circuit are voltage level E=200mV, period T = 10ms, time of the impulse t_i = 8ms, resistance R = 1kΩ, capacitance C = 10μF. Will be calculated the capacitive voltage at time t = 46ms using analytical formulas. Will be analyzed the circuit by PSPICE simulation for the capacitive voltage $u_c(t)$. Will be compared analytical results with simulation results.

Analytical Solution 3

First next values are calculated

$\tau = RC = 2.10^3.13.10^{-6} = 0.010,\ \ T/\tau = 1.0,\ \ 5T/\tau = 5.0,\ \ t_i/\tau = 0.8,\ \ t'/\tau = 0.7$.

After that these data are substituted in expression (21) and the capacitive voltage is obtained

$$u_C(0.006,5) = 0.2 + 0.2\left\{\frac{1-e^{0.7}}{1-e^{-1.0}}e^{-5} - \frac{1-e^{-0.2}}{1-e^{-1.0}}\right\}e^{-0.7} = 0.1697V$$

Simulations 3

The capacitive voltage $u_C(t)$ is simulated by PSPICE and obtained graph is shown on Fig.4. The capacitive voltage at the time t = 46*ms*, computed by simulation is $u_C(46ms)$ = 169.9*mV*. The used time t = 46*ms* refers to the 5 period (n = 5) of the impulses and the corresponding time is $t' = 6ms$.

Over again, simulation results and analytical solution has good match.

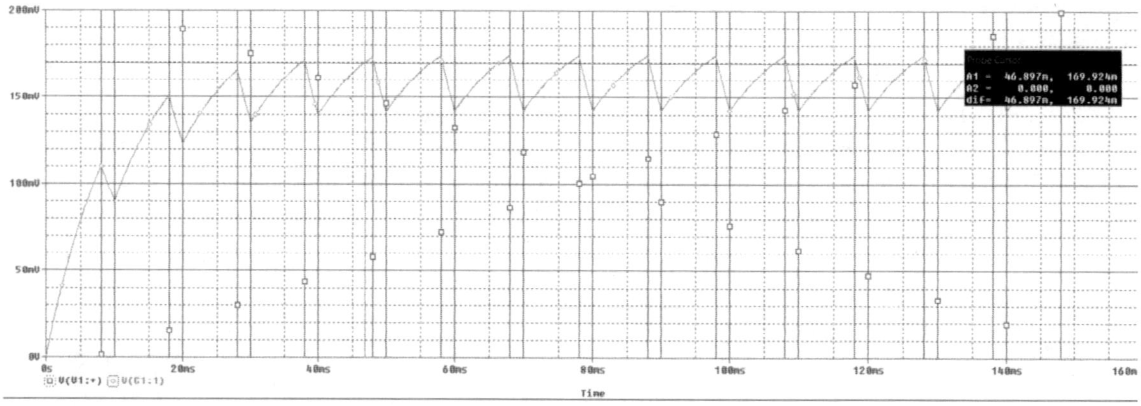

Fig. 4. A capacitive voltage obtained by simulation at $t_i = 0.8ms$

In this case, the transient process is stable and short, and the level of the capacitive voltage is high enough.

5. Conclusions

The problems of discrete analysis of the transients and stationary process in electric circuits supplyed by impulse voltage sources are studied. An elementary RC circuit powered by rectangular pulses is considered.

Z-transformation is using as approach for discrete analysis of circuits, supplied by rectangular pulses. In this way, z-transformation is used to find transfer function, discrete impulse function and capacitive voltage.

The obtained analytical dependences enable the determination of the capacitive voltage at any moment of time. Two analytical expressions are obtained for the two cases of computing time - during the pulse and during the pause.

Investigations were carried out at different parameters of the triangular pulses.

In each case, calculations are performed on the analytical dependences for the capacitive voltage and computer simulations. In each case, good matches between the analytical solutions and the simulations are obtained.

The method proposed in the present work allows, together with the transient process, to investigate the corresponding stationary process. Cases of stable and unstable transients are shown.

It is shown that for certain times of the pulse, an unstable transient process occurs, and in the steady-state mode the level of the output capacitive voltage is unsatisfactorily low.

The authors would like to thank the Research and Development Sector at the Technical University of Sofia for the financial support.

References

[1] Brandisky K, Georgiev Zh, Mladenov V, Stancheva R 2008 *Theoretical electrical engineering* – vol. **2**, King, Sofia (in Bulgarian).

[2] Alexander C, M. Sadiku M 2013, *Fundamentals of Electric Circuits* 5th Edition, McGraw-Hill.

[3] Demirchan K S, Neyman L R, Korovkin N V, Chechurin V L 2004 *Theoretical electrical engineering* rev. 4, vol.2, Piter, Sankt Petersburg.

[4] Shopov Y, Petrakieva S 2019 Overvotages Protection in Electrical Devices and Transient Analysis by Duhamel's Integral, *11th Electrical Engineering Faculty Conference BulEF'2019, 11-14.09.2019, Varna, Bulgaria*, , doi: 10.1109/BulEF48056.2019.9030791.

[5] Georgiev Zh, Trushev I, Chervenkov A 2017 Transient and steady-state analysis of electrical circuits supplied by sequence of impulses by Z-transformation *Proceedings of Technical University of Sofia* vol. 66, issue 2, pp. 373-382.
[6] Bainov D, Simeonov P 1989 *Systems with Impulse Effects: Stability, Theory and Applications* Academy Press.
[7] Palani S 2021 The z-Transform Analysis of Discrete Time Signals and Systems *Signals and Systems,* doi:10.1007/978-3-030-75742-7_9.
[8] Vich R 1987 *Z-Transform Theory and Applications*", D. Reidel Publishing Company.
[9] Dobrucký B, Štefanec P, Koňarik R, Chernoyarov O V 2016 Transient analysis in electrical circuits using Z-transformation *ELEKTRO 2016, IEEE*, pp. 131-136.
[10] Rubeena AA, Paulose S, Thomas M, Joy J 2022 PWM control strategies for Switched-Capacitor inverters *Materials Today: Proceedings*, vol. 58, part 1, pp. 516-522.
[11] Bourgeade A, Ghanes M, Barbot Jean-Pierre., Abdelkader B, Maurice F, Boisliveau R 2022 New PWM inverter control based on optimal pulse pattern computation without phases symmetry constraints *Control Engineering Practice*, vol. 132, 105390, https://doi.org/10.1016/j.conengprac.2022.105390.
[12] Bushra E, Zeb K, Ahmad I, Khalid M 2024 A comprehensive review on recent trends and future prospects of PWM techniques for harmonic suppression in renewable energies based power converters *Results in Engineering*, vol. 22, 2024, 102213, ISSN 2590-1230, https://doi.org/10.1016/j.rineng.2024.102213.
[13] Zamiri E, Sanchez A, María-Sofía Martínez-García, Angel de Castro 2022 Analysis of the aliasing effect caused in hardware-in-the-loop when reading PWM inputs of power converters *International Journal of Electrical Power & Energy Systems*, vol.136, 2022, 107678, https://doi.org/10.1016/j.ijepes.2021.107678.

IOP Conf. Series: Materials Science and Engineering **1317** (2024) 012010 doi:10.1088/1757-899X/1317/1/012010

The Two-Wheeled Robot: Experimental Evaluation of Two Controllers

Tsonyo N. Slavov[1], Jordan K. Kralev[1] and Petko H. Petkov[2]

[1]Department of Systems and Control, Technical University of Sofia, 1157 Sofia, Bulgaria

[2]Bulgarian Academy of Sciences, 1040 Sofia, Bulgaria

E-mail: ts_slavov@tu-sofia.bg

Abstract. In this paper we present the theoretical and experimental comparison of two digital controllers intended for stabilization of a two-wheeled robot. The properties of an LQG controller and an LQR controller with H_∞ filter are evaluated. Using the μ-analysis, we show that both controllers ensure robust stability of the closed-loop system in the presence of unstructured uncertainty. The results from the closed-loop system experimental evaluation confirm the precise work of both controllers.

1 Introduction

The control of a statically unstable Two-Wheeled Robot (TWR) is a subject of several research papers recently, see for instance Do et al. (2024); Kralev et al. (2016); Mishra and Bansal (2024); Petkov et al. (2018); Slavov et al. (2021, 2023). Due to the inherent instability of the plant, the derivation of a sufficiently accurate dynamics TWR model with uncertainties in the plant, sensors and actuators is a difficult task. For this aim we implement an efficient closed-loop Single-Input Multi-Output (SIMO) identification method presented in Ljung (1999, 2022), which allows to obtain a low order linear small deviations model. The determination of an uncertain robot model is performed from experiment results as described in Hassan, Moustafa and Moness (2024); Petkov et al. (2018); Slavov et al. (2021, 2023). Typically, as a result of such modeling, one obtain a robot description in the form of Linear Fractional Transformation (LFT) that reflects the plant uncertainty. This LFT is used in the controller design and in the time-domain and frequency domain analysis of the closed-loop system.

In this paper, we present the experimental and theoretical evaluation of two linear quadratic controllers for the TWR. In the controller design, we implement the SIMO model with uncertainty determined by the aid of the identification experiment. For this aim, we use the model derived in Slavov et al. (2021), which was previously used in the LQG and μ controllers of the robot Slavov et al. (2021, 2023). We compare the closed-loop frequency responses of a discrete-time LQG controller and a LQR controller with H_∞ filter. By implementing the μ-analysis Gu et al. (2013), we show that both controllers ensure robust stability of the closed-loop system despite of the presence of unstructured uncertainty. The controllers are embedded in the microcontroller Raspberry Pi 4B and the closed-loop system experimental evaluation confirms the accurate work of both controllers.

The paper contains four sections. In section 2 we provide a brief information about the TWR under investigation referring the reader to the paper Petkov et al. (2018) for a more detailed description. We utilize a 5th order Single-Input Two-Output (SITO) robot model according to the description given in Slavov et al. (2021). Instead of the LQG design, done in Slavov et al. (2021), in section 3 we perform the design of an 8th order LQR controller with H_∞ filter. We perform an evaluation of the frequency responses and robustness properties of the closed-loop system for this controller and the the LQG controller with Kalman filter. Finally, in section 4 we present the results obtained from the experimental comparison of the two controllers.

IOP Conf. Series: Materials Science and Engineering **1317** (2024) 012010 doi:10.1088/1757-899X/1317/1/012010

2 The Uncertain Model

The deviation of the robot position from the vertical position is characterized by the tilt angle ϕ and its horizontal motion can be described by the angle $\theta = (\theta_L + \theta_R)/2$ describing the average wheels motion, where θ_L and θ_R denote the angle position of the left and right wheel angles. The motion of the robot motion with respect to the vertical axis is steered by a separate controller and is not considered here. The robot control system consists of two servo motors for wheels actuation, MEMS sensor measuring the angular velocity $\dot{\phi}$ of the robot body, quadrature encoders measuring the wheels position, and a digital controller in which a discrete real-time stabilization algorithm is implemented. The sampling period of the control action is equal to $T_s = 0.005$ s. The balancing of the robot is achieved by rotating the wheels in the appropriate direction. The calculation of the control inputs to both DC drive motors is done in single precision based on signals from the MEMS sensor that measures the rate $\dot{\phi}$ (an estimate of the angle ϕ is obtained by integrating $\dot{\phi}$) and signals from the rotary encoders measuring the wheels rotation angle θ. The control of the DC brushed motors is realized by Pulse Width Modulated (PWM) signals.

In this paper, the wheels dynamics and the vertical robot motion are described by an estimated SIMO state-space model whose input is the control action u and outputs are the wheels angular rate $\dot{\theta}$ and the tilt angle ϕ of the robot body. As mentioned previously, this model was already used in Slavov et al. (2021, 2023) for other types of controller design. A peculiarity of the model is that as a result from the identification, it has a large number of structure uncertainties which makes its implementation unfeasible. Hence, it is appropriate to obtain transfer function models $G_{\dot{\theta}u}$ and $G_{\phi u}$ that contain complex (unstructured) uncertainties Gu et al. (2013). Note that this is associated with the introduction of some conservatism in the model. Using the parameter confidence intervals, obtained by the identification, we obtain the maximum relative deviations from the nominal model in the frequency domain. Assuming the presence of output unstructured uncertainties, we can approximate these deviations by using an optimization procedure of shaping filters. These filters are represented by 3rd order transfer functions. As a result of the identification procedure, the nominal 5th order discrete-time state space model of the robot is obtained as

$$\begin{aligned} x_{k+1} &= Ax_k + Bu_k + K_e e_k, \\ y_k &= Cx_k + Du_k + e_k, \end{aligned} \tag{1}$$

where x_k, u_k and $y_k = [\dot{\theta},\ \phi]^T$ and $e_k = [e_{\dot{\theta}},\ e_\phi]^T$, are the state, input, output and residue error vectors. The model matrices A, B, C, D, K_e of the observable canonical form are

$$A = \begin{bmatrix} 0 & 1 & 0 & 0 & 0 \\ -7.0976 \cdot 10^{-1} & 1.6834 \cdot 10^{0} & -7.9018 \cdot 10^{-1} & 9.9694 \cdot 10^{-2} & 7.6250 \cdot 10^{-1} \\ 0 & 0 & 0 & 1 & 0 \\ 0 & 0 & 0 & 0 & 1 \\ -2.9747 \cdot 10^{-4} & 2.8183 \cdot 10^{-4} & 5.6080 \cdot 10^{-1} & -2.1149 \cdot 10^{0} & 2.5546 \cdot 10^{0} \end{bmatrix},$$

$$B = \begin{bmatrix} -3.8502 \cdot 10^{-3} \\ 1.6537 \cdot 10^{-1} \\ -1.2123 \cdot 10^{-5} \\ 5.8352 \cdot 10^{-5} \\ 2.2311 \cdot 10^{-4} \end{bmatrix}, \tag{2}$$

$$K_e = \begin{bmatrix} 8.3952 \cdot 10^{-1} & -5.1882 \cdot 10^{-1} \\ 7.4844 \cdot 10^{-1} & 5.9226 \cdot 10^{-1} \\ 1.6353 \cdot 10^{-4} & 1.8714 \cdot 10^{0} \\ 5.4093 \cdot 10^{-4} & 2.7881 \cdot 10^{0} \\ 1.0845 \cdot 10^{-3} & 3.3755 \cdot 10^{0} \end{bmatrix},$$

$$C = \begin{bmatrix} 1 & 0 & 0 & 0 & 0 \\ 0 & 0 & 1 & 0 & 0 \end{bmatrix},$$

$$D = \begin{bmatrix} 0 \\ 0 \end{bmatrix}.$$

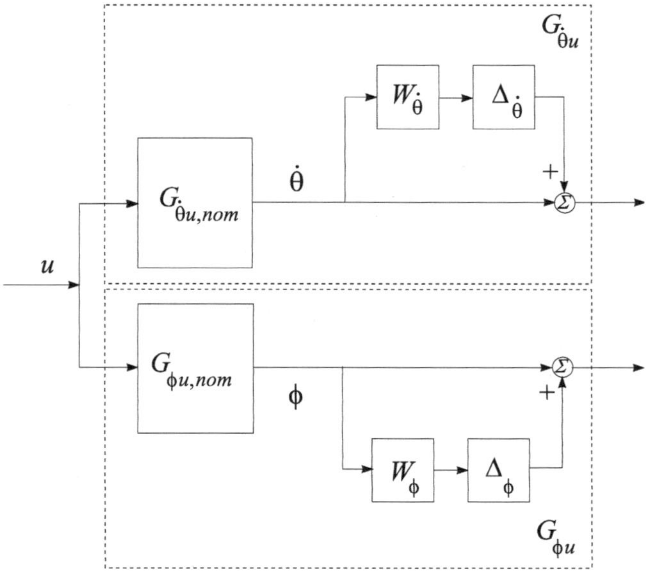

Figure 1: Uncertain plant model

The SITO uncertain model involving unstructured uncertainties is described by

$$G_{\dot\theta u}(z) = (1 + W_{\dot\theta}(z)\Delta_{\dot\theta})G_{\dot\theta u,nom}(z), \quad (3)$$

$$G_{\phi u}(z) = (1 + W_{\phi}(z)\Delta_{\phi})G_{\phi u,nom}(z), \quad (4)$$

where $G_{\dot\theta u,nom}(z)$, $G_{\phi u,nom}(z)$ are the transfer functions of the nominal model derived from the state-space description (1), (2),

$$W_{\dot\theta}(z) = \frac{0.27680z^3 - 0.65150z^2 + 0.48620z - 0.11120}{z^3 - 1.66000z^2 + 0.50270z + 0.16010},$$

$$W_{\phi}(z) = \frac{0.83430z^3 - 1.82600z^2 + 1.25700z - 0.26200}{z^3 - 1.11800z^2 + 0.21280z - 0.08790}$$

are the 3rd order shaping filters that represent the output multiplicative uncertainty and $\Delta_{\dot\theta}, \Delta_{\phi}$ are the output multiplicative uncertainties that satisfy

$$|\Delta_{\dot\theta}| < 1, \;\; |\Delta_{\dot\phi}| < 1.$$

In Fig. 1 we show the model of the TWR involving output multiplicative uncertainties. The order of the uncertain plant model is equal to 11.
In Fig. 2 and Fig. 3, we present families of frequency response plots corresponding to the models $G_{\dot\theta u}$ and $G_{\phi u}$.

3 Design of LQR Controller With H_∞ Filter

In Fig. 4 we show the closed-loop system block-diagram implemented in the design of the LQR controller with H_∞ filter, where r_θ is the reference signal for the wheels motion. In the design, we look for a controller K for which the closed-loop system achieves acceptable closed-loop robust performance under the modelled unstructured uncertainties Δ_ϕ and $\Delta_{\dot\theta}$. A better position accuracy can be achieved by adding the integral of the tracking error $err = r_\theta - \theta$ as additional controller input. The real-time control signal to the plant is realized by a microcontroller with sampling period $T_s = 0.005$ s. That is why we perform the design of the discrete-time controller at sampling frequency $f_s = 200$ Hz.

IOP Conf. Series: Materials Science and Engineering **1317** (2024) 012010 doi:10.1088/1757-899X/1317/1/012010

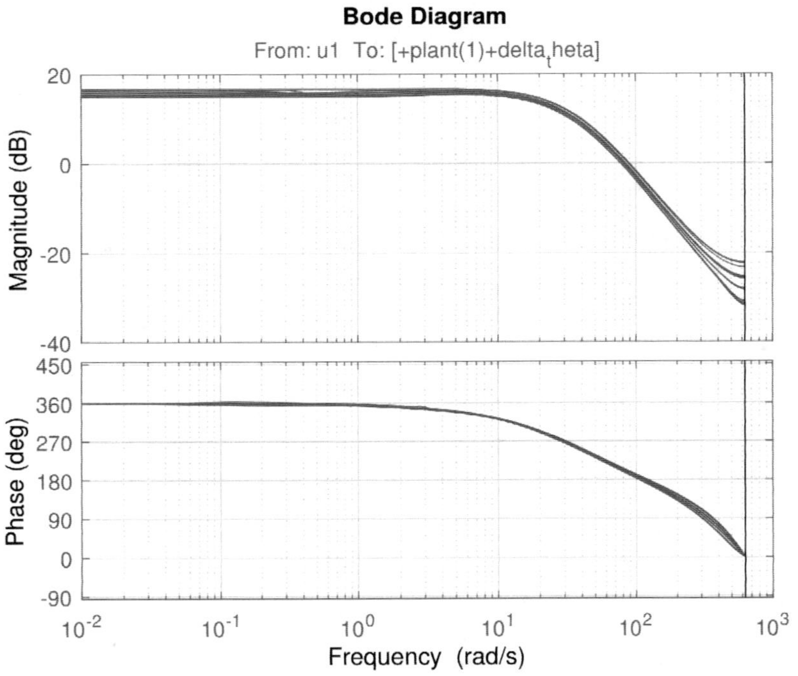

Figure 2: The frequency responses corresponding to the uncertain transfer function $G_{\dot{\theta}u}$

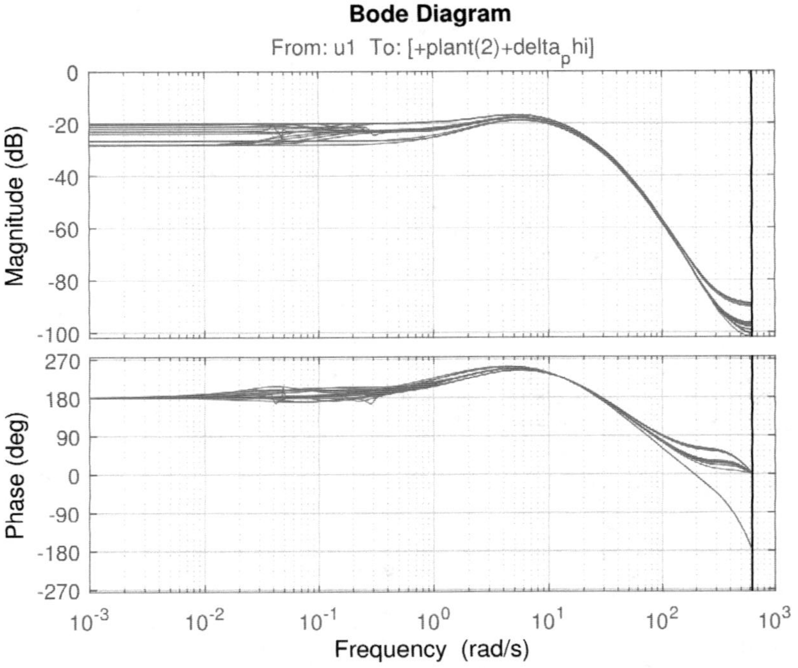

Figure 3: The frequency responses corresponding to the uncertain transfer function $G_{\phi u}$

IOP Conf. Series: Materials Science and Engineering **1317** (2024) 012010 doi:10.1088/1757-899X/1317/1/012010

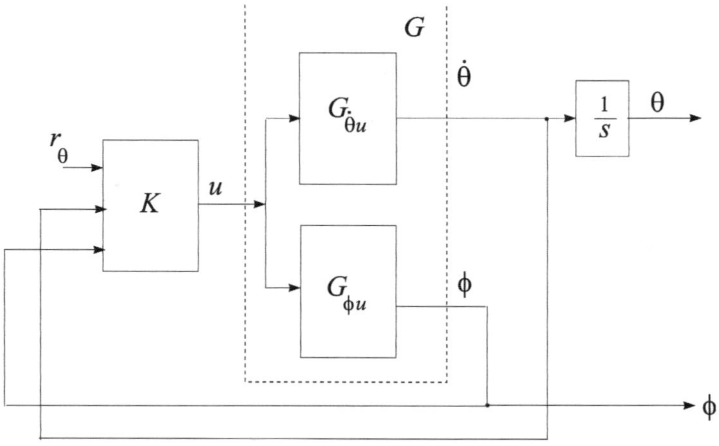

Figure 4: Closed-loop system structure

3.1 LQR Design

The design of the LQG controller for the TWR is considered in detail in SlaPSK2021. In the LQR controller design, we use the 5th order state space model

$$
\begin{aligned}
x_{k+1} &= Ax_k + Bu_k + Gv_k, && (5)\\
y_k &= Cx_k + Hv_k, && (6)
\end{aligned}
$$

where the matrices A, B, C are given in (2) and $v = [v_1,\ v_2]^T$ is a white Gaussian vector noise. Such a noise is a consequence of the unbiased model parameter estimates. The matrices G and H are determined as

$$
G = \begin{bmatrix} 7.468 & -0.006557 \\ 6.659 & 0.007485 \\ 0.003507 & 0.02365 \\ 0.00787 & 0.03524 \\ 0.01335 & 0.04266 \end{bmatrix}, H = \begin{bmatrix} 8.897 & 0 \\ 0.001097 & 0.01264 \end{bmatrix}.
$$

To ensure accurate tracking of the reference signal r_Θ corresponding to the desired wheels position, the model (5) is extended with the additional states

$$
\begin{aligned}
x_{\Theta,k+1} &= x_{\Theta,k} + T_s\dot{\Theta}_k = x_{\Theta,k} + T_sC_1x_k + T_sH_1v_k, && (7)\\
x_{\text{int}\Theta,k+1} &= x_{\text{int}\Theta,k} + T_s(r_\Theta - \Theta_k)x_{\text{int}\Theta,k} + T_sr_\Theta - T_sx_{\Theta,k} && (8)\\
&\quad -T_sx_{\Theta,k} - T_s^2C_1x_k - T_s^2H_1v_k, \\
x_{\text{int}\phi,k+1} &= x_{\text{int}\phi,k} - T_s\phi_k = x_{\text{int}\phi,k}T_sC_2x_k - T_sH_2v_k, && (9)
\end{aligned}
$$

where the unmeasurable wheel position $x_{\Theta,k}$ is obtained by numerical integration of $\dot{\Theta}$. The variable $x_{\text{int}\Theta}$ is the integral of the wheel position error, $x_{\text{int}\phi}$ is the integral of the body tilt angle error and C_1, H_1, C_2, H_2 are the corresponding rows of the matrices C and H. Combining equations (5) - (9), we obtain the extended plant equations

$$
\begin{aligned}
\bar{x}_{k+1} &= \bar{A}\bar{x}_k + \bar{B}u_k + \bar{G}v_k, && (10)\\
\bar{y}_k &= \bar{C}\bar{x}_k + \bar{v}_k,
\end{aligned}
$$

where $\bar{x}_k = [x_{\text{int}\Theta}, x_{\text{int}\phi}, x_\Theta, x^T]^T$ is the 8th order extended state vector and $\bar{y}_k = [\Theta,\ \dot{\Theta},\ \phi]^T$ is the extended output vector.

In the design of the LQR controller, we use the deterministic part

$$
\bar{x}_{k+1} = \bar{A}\bar{x}_k + \bar{B}u_k \qquad (11)
$$

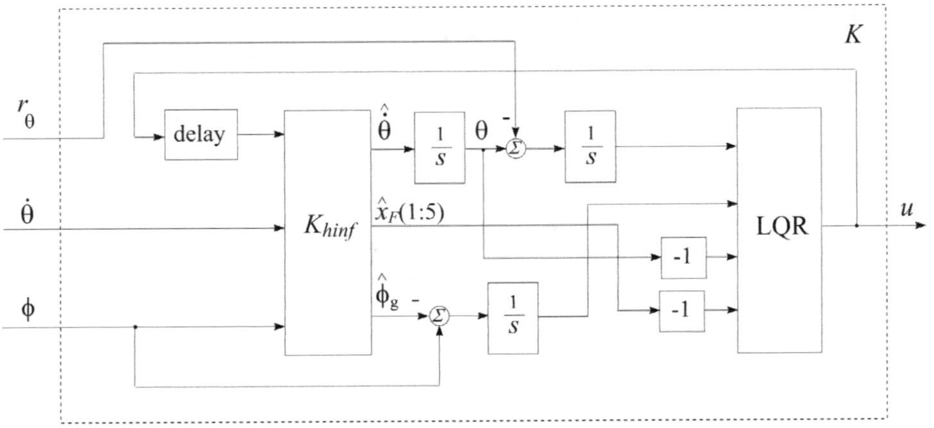

Figure 5: Block-diagram of the LQR controller with H_∞ filter

of the extended model (10). The optimum performance index is chosen as usual in the form

$$J = \sum_{k=0}^{\infty} \bar{x}_k^T Q \bar{x}_k + u_k^T T u_k, \tag{12}$$

where the weighting matrices Q and R are taken as

$$Q = \mathrm{diag}(Q_{\mathrm{int}\Theta}, Q_{\mathrm{int}\phi}, Q_\Theta, C^T Q_{\dot{\Theta}\phi} C), R = 100,$$

with $Q_{\mathrm{int}\Theta} = 30$, $Q_{\mathrm{int}\phi} = 25$, $Q_\Theta = 1$ and $Q_{\dot{\Theta}\phi} = \mathrm{diag}(12,\ 5)$. The discrete-time optimal control law minimizing (13), is

$$u_k = -K\bar{x}_k, \tag{13}$$

where

$$K = (R + \bar{B}^T P \bar{B})^{-1} \bar{B}^T P \bar{A} \tag{14}$$

and the positive definite matrix P is the solution of the 8th order Riccati equation

$$\bar{A}^T P \bar{A} - P - \bar{A}^T P \bar{B}(R + \bar{B}^T P \bar{B})^{-1} \bar{B}^T P \bar{A} + Q = 0. \tag{15}$$

The LQR controller design is done by `dare` function from the Control System Toolbox of MATLAB®.

3.2 H_∞ filter design
Consider the discrete time invariant plant

$$\begin{aligned} x_{f,k+1} &= A_f x_{f,k} + B_f u_k + w_k, \\ y_{f,k} &= C_f x_{f,k} + v_k, \end{aligned} \tag{16}$$

where w_k and v_k are the state and output disturbances. The disturbances can be random noises with known or unknown probability characteristics but also can be unmeasured signals. Define the objective function as

$$J_0 = \frac{\sum_{k=0}^{N-1} (x_{f,k} - \hat{x}_{f,k})^T S_f (x_{f,k} - \hat{x}_{f,k})}{(w_k^T Q_f^{-1} w_k + v_k^T R_f^{-1} v_k)}, \tag{17}$$

where Q_f, R_f and S_f are weighting matrices that are symmetric and positive definite. The aim is to obtain an estimate $\hat{x}_{f,k}$ of the state $x_{f,k}$ which minimizes (17). Since the direct minimization of (17) is difficult, we set a bound on the performance and look for a solution that satisfies it, i.e., we look for an estimate $\hat{x}_{f,k}$ such that

$$J_0 < \gamma, \tag{18}$$

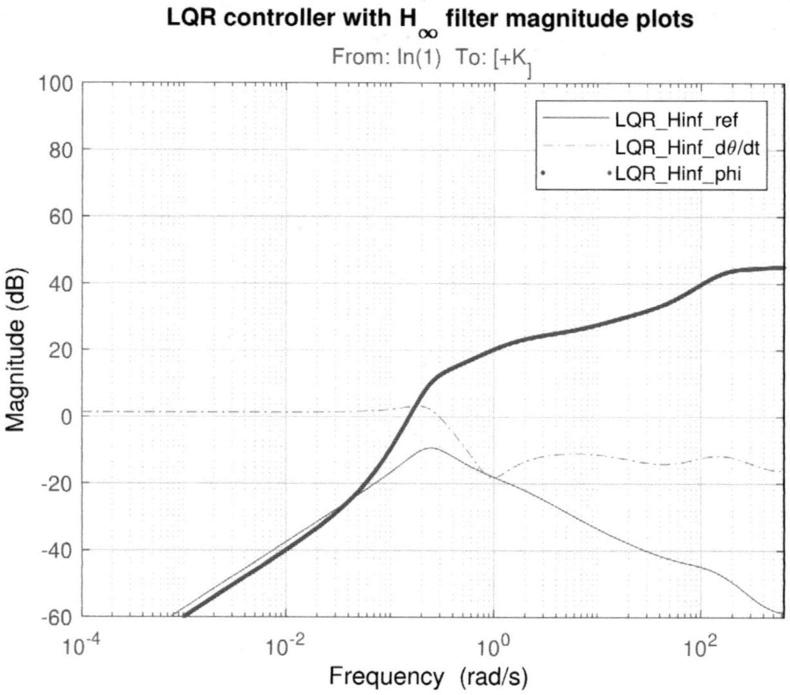

Figure 6: Magnitude plots of the LQR controller with H_∞ filter with respect to the separate inputs

where γ is a bound on the estimate performance. From (17) and (18) we obtain that

$$J = \sum_{k=0}^{N-1} (x_{f,k} - \hat{x}_{f,k})^T S_f (x_{f,k} - \hat{x}_{f,k}) - \gamma (w_k^T Q_f^{-1} w_k + v_k^T R_f^{-1} v_k) < 0.$$

The solution of the optimization problem

$$J^* = \min_{\hat{x}_{f,k}} \max_{w_k, v_k} J \tag{19}$$

yields an optimal estimate $\hat{x}_{f,k}$ satisfying (18) which is obtained by the filter

$$\hat{x}_{f,k+1} = A_f \hat{x}_{f,k} + B_f u_k + A_f K_f (y_k - C_f \hat{x}_{f,k}), \tag{20}$$

The gain matrix K_f of the H_∞ filter is determined by

$$K_f = P_f (I - \frac{1}{\gamma} S_f P_f + C_f^T R_f^{-1} C_f P_f)^{-1} C_f^T R_f^{-1} \tag{21}$$

and the positive definite matrix P_f is the unique solution of the Riccati equation

$$A_f P_f A_f^T - P_f + Q_f - A_f P_f [(C_f^T R_f^{-1} C_f - \frac{1}{\gamma} S_f)^{-1} + P_f]^{-1} P_f A_f^T = 0. \tag{22}$$

For (20), (21) and (22) to be solutions of (19), it is necessary to satisfy the condition

$$P_f (I - \frac{1}{\gamma} S_f P_f + C_f^T R_f^{-1} C_f P_f)^{-1} > 0. \tag{23}$$

The block-diagram of the LQR controller with H_∞ filter is shown in Fig. 5. The block with the name "delay"reflects the presence of one sample delay in the control action calculation.

The state vector x_k of the TWR is unmeasurable, so that the control (3) is obtained from

$$u_k = -K_{\dot{\phi}_i}\hat{x}_{\dot{\phi}_i,k} - K_{x_i}\hat{x}_{i,k} - K_{\Theta}\hat{\Theta}_k - K_{\dot{\phi}}\hat{x}_{\dot{\phi},k} - K_{\dot{\Theta}}\hat{x}_{\dot{\Theta},k}, \tag{24}$$

where $\hat{x}_{\dot{\phi}}, \hat{x}_{\dot{\Theta}}$ denote the estimates of $x_{\dot{\phi}}$, $x_{\dot{\Theta}}$ and $\hat{x}_{\dot{\phi}_i}$, $\hat{x}_i$,

$$\begin{aligned} \hat{x}_{\dot{\phi}_i,k+1} &= \hat{x}_{\dot{\phi}_i,k} - T_s\hat{\dot{\phi}}_k,\ \hat{\dot{\phi}}_k = C_{\dot{\phi}}\hat{x}_{\dot{\phi},k}, \\ \hat{x}_{i,k+1} &= \hat{x}_{i,k} + T_s(r_{\Theta,k} - \hat{\Theta}_k), \end{aligned} \tag{25}$$

denote estimates of $x_{\dot{\phi}_k}$ and x_{ik}. The estimates $\hat{x}_{\dot{\phi},k}, \hat{x}_{\dot{\Theta},k}$ and $\hat{\Theta}_k$ are obtained by the H_∞ filter. The gyro sensor used to measure $\dot{\phi}_k$ is characterized by the noise ϕ_g which is a solution of the equation

$$\dot{\phi}_{g,k+1} = \phi_{g,k} + G_g v_{g,k}, \tag{26}$$

where $v_{g,k}$ is a discrete white gaussian noise and $G_g = 0.002$ is a coefficient that is experimentally determined to obtain an accurate estimate of the angular velocity $\dot{\phi}_k$. Combining the equations (1) and (26), we obtain a model of 6th order in the form (16), where

$$A_f = \begin{bmatrix} 1 & T_sC_1 \\ 0 & A \end{bmatrix}, B_f = [\ B \quad 0\ ], C_f = \begin{bmatrix} C & \begin{bmatrix} 0 \\ 1 \end{bmatrix} \end{bmatrix}, G_f = \begin{bmatrix} G & 0 \\ 0 & G_g \end{bmatrix}.$$

The solution of the Riccati equation (22) is obtained for $Q_f = G_f I_3 G_f^T$, $R_f = 10^{-6} I_2$ and $S_f = I_6$. To obtain a positive definite solution of the equation, the parameter γ is taken equal to 40. The H_∞ filter designed is of 6th order.
The frequency responses corresponding to the transfer functions $K_{\int err}$, K_θ, K_ϕ and $K_{\dot{\theta}}$ associated with the controller inputs, are shown in Fig. 6.

3.3 Comparison of the System Properties with LQG Controller and LQR Controller with H_∞ Filter

In this section, we compare the frequency domain properties of the closed-loop systems with LQG controller designed in Slavov et al. (2021) with the closed-loop properties of the LQR controller with H_∞ filter whose design was presented in the previous section.
In Figure 7 we show the Bode plot of the closed-loop system for the two controllers. We see that for both controllers the closed-loop system bandwidth is around 1 rad/s.
The singular value plot corresponding to the tracking error transfer function, is shown in Figure 8. The two controllers ensure the same tracking error attenuation up to the frequency $2 \cdot 10^{-1}$ rad/s.
The singular value plot of the input sensitivity function, given in Figure 9, shows that the two controllers attenuate in the same way the noises and disturbances in the control signal.
In Figure 10, we give the upper bound plots of the structured singular value μ with respect to the closed-loop robust stability for the two controllers Gu et al. (2013). The value of μ is determined by the `robuststab` function from Robust Control Toolbox®3 Balas et al. (2022). Both controllers ensure closed-loop robust stability with respect to the unstructured uncertainties.

4 Experimental Results

The experimental comparison of the two controllers designed is performed as described in Petkov et al. (2018) by the aid of the Simulink Coder®tool SimCoder (2021). The code of control algorithm is generated by using the Simulink® model. It is implemented in the digital controller as described in Raspberry Pi 4 (2021). The communication between the computer and the robot controller is performed via conventional TCP/IP protocol.
The simulation of the TWR control is done for 300 s (60000 samples). The maximum computational time is about $5 \cdot 10^{-5}$ s for both control actions which is much smaller than the sampling period $T_s = 0.005$ s.
In Fig. 11, we depict the dynamics of the wheel angle for the two controller algorithms. It is seen that for both control systems have good static performance.
The oscillations that are presented in the tilt body angle are negligible, as seen in Fig. 12.
The experimentally obtained control actions are presented in Fig. 13. As a control action we consider the duty PWM coefficient that is the input to the DC motors. The maximum control action for both controllers is smaller than 5 units, which is so far from the upper allowable bound of 128 units. We note the relatively strong effect of the noise on the control action as a results of the form of the input sensitivity function (Fig. 9).

IOP Conf. Series: Materials Science and Engineering **1317** (2024) 012010 doi:10.1088/1757-899X/1317/1/012010

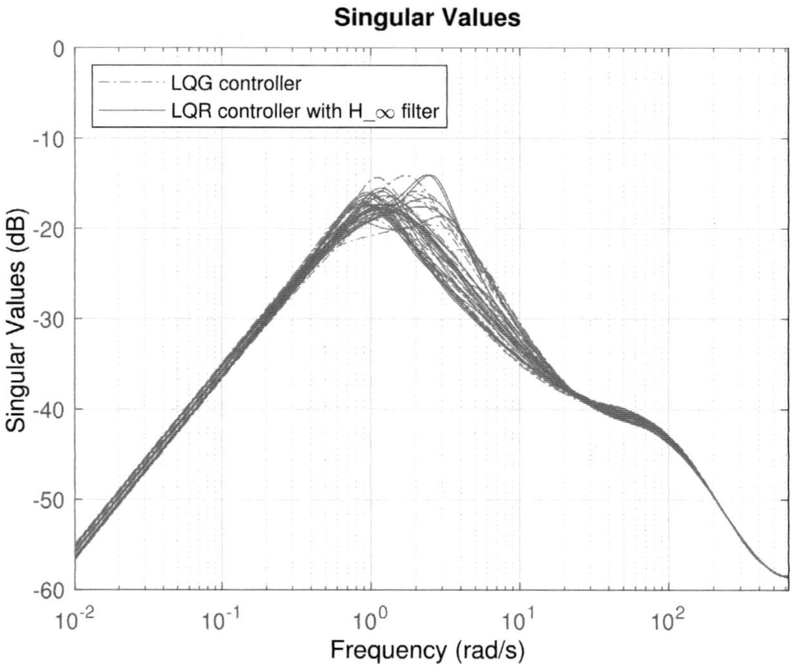

Figure 7: Bode plot of the uncertain closed-loop system

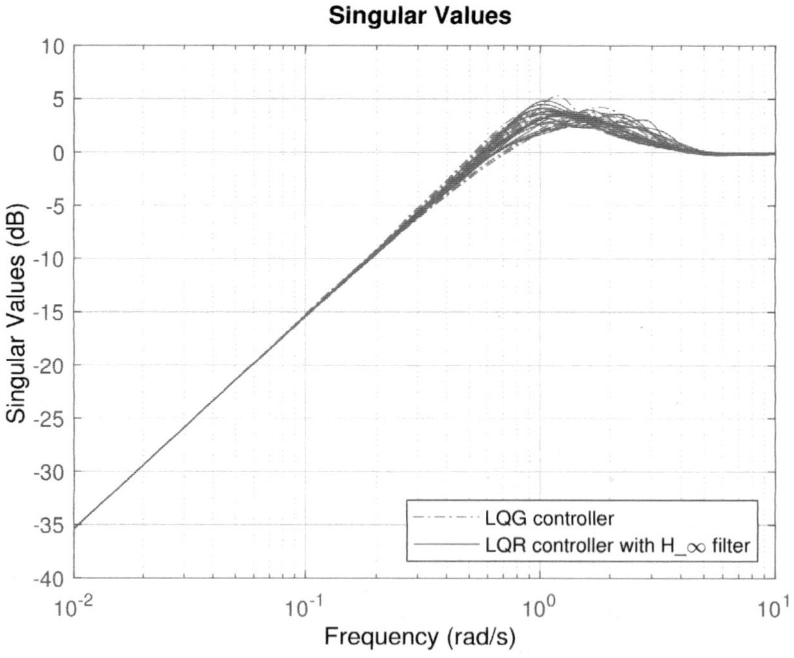

Figure 8: Singular value plot of the output sensitivity function $S_o G_{r_\theta}$

IOP Conf. Series: Materials Science and Engineering **1317** (2024) 012010 doi:10.1088/1757-899X/1317/1/012010

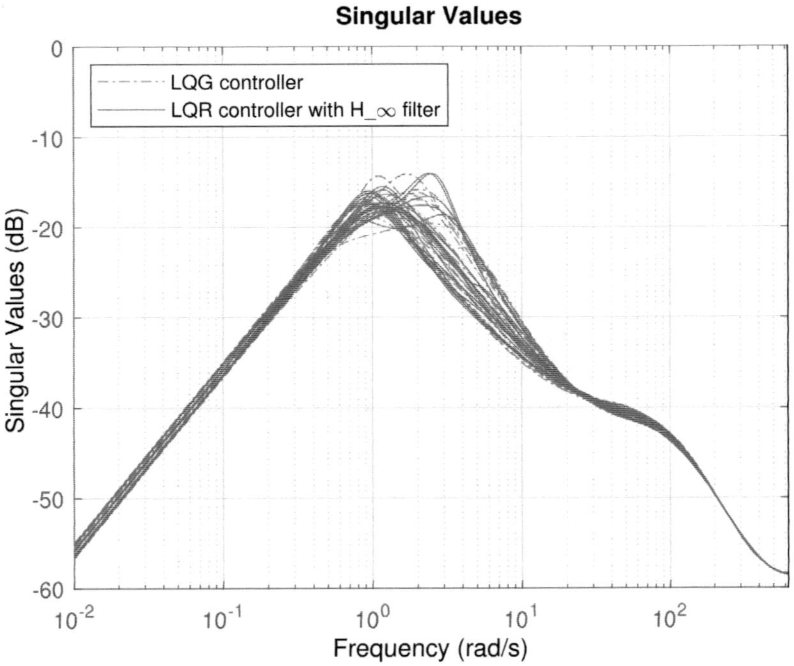

Figure 9: Singular value plot of the input sensitivity function $S_i K G_{r_\theta}$

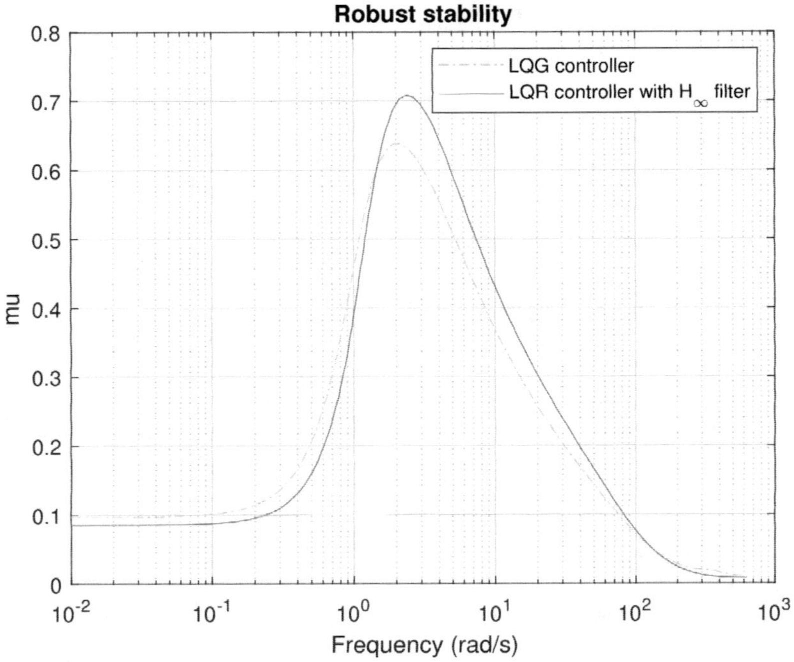

Figure 10: Robust stability of the closed-loop system

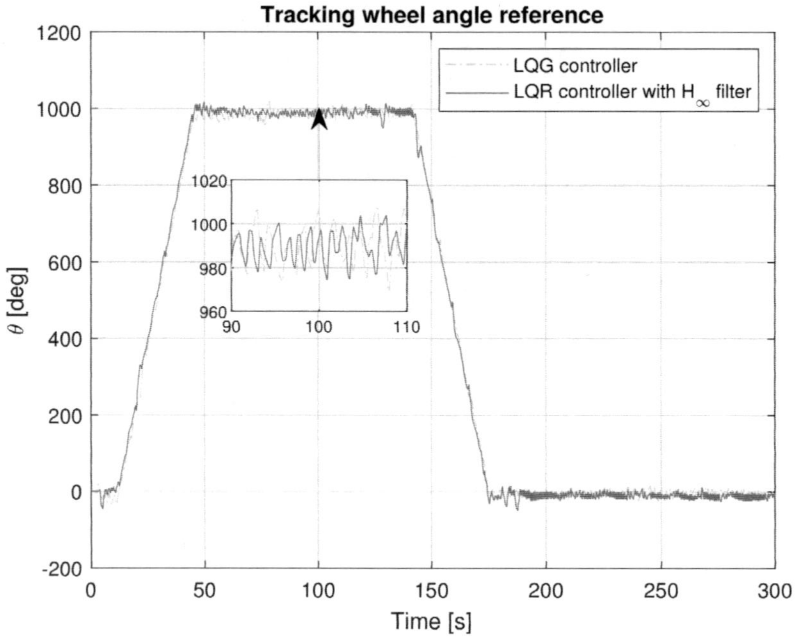

Figure 11: Tracking wheel angle reference

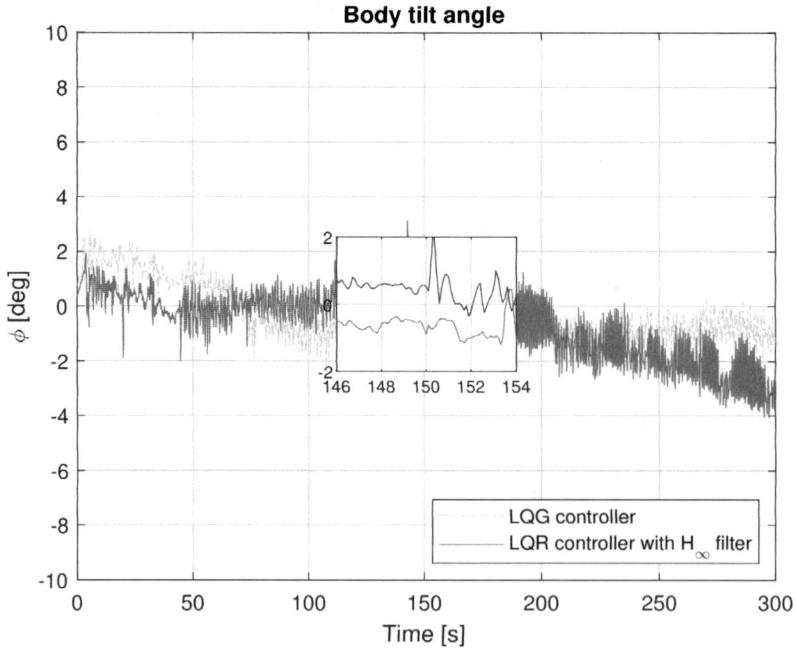

Figure 12: Tilt body angle

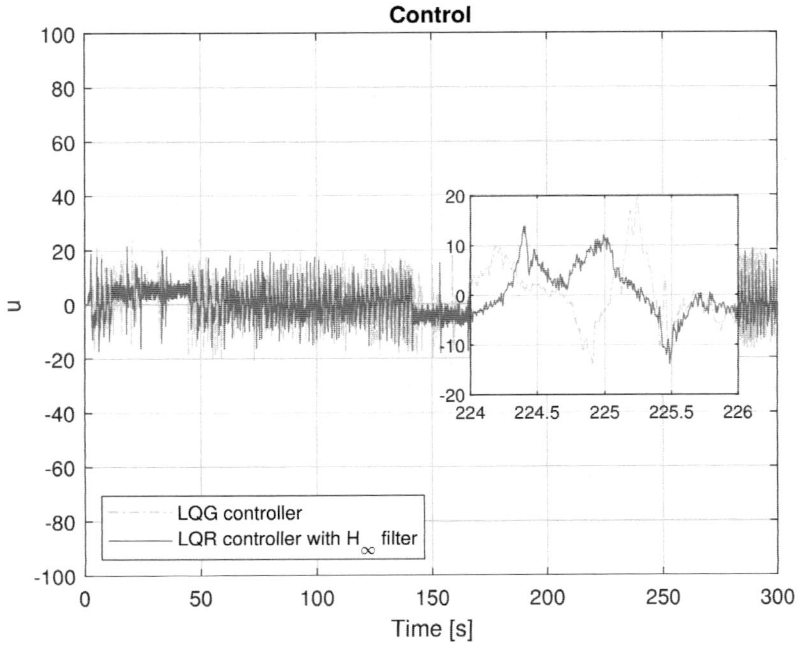

Figure 13: Control action

5 Conclusions

The comparison of the closed-loop performance of the discrete-time LQG controller with the LQR controller with H_∞ filter shows that the two controllers have nearly the same properties in the time domain and frequency domain. Also, their robust stability properties are almost the same. Both controllers can be used in practice to ensure accurate position tracking and stabilization.

Acknowledgment

This study is financed by the European Union-NextGenerationEU, through the National Recovery and Resilience Plan of the Republic of Bulgaria, project № BG-RRP-2.004-0005.

References

Balas, G., Chiang, R., Packard, A., and Safonov, M. (2022). *Robust Control Toolbox® 3 User's Guide.* Natick, MA: The MathWorks, Inc.
`https://www.mathworks.com/help/pdf_doc/robust/robust_ug.pdf`

Do, Q.-T., Tran, V.T., Ngo, M.T. et al. (2024). Modeling and optimal cntrol for two-wheeled self-balancing robot. *J. Fuzzy Syst. & Control, 2, 22–28.* DOI: 10.59247/jfsc.v2i1.162

Gu, D.-W., Petkov, P.Hr., and Konstantinov, M.M.(2013). *Robust Control Design with MATLAB®.* Springer-Verlag, London, 2nd ed. ISBN: 978-1-4471-4681-0

Hassan, M., Moustafa, A.M. and Moness, M. (2024). Modeling and Parameters Estimation of a Self-Balancing Two-Wheeled Vehicle. *J. Advanced Engng. Trends, 43, 373–381.* DOI: 10.21608/jaet.2022.144785.1197

Kralev, J., Slavov, Ts., and Petkov, P. (2016). Design and experimental evaluation of robust controllers for a TWR. *Int. J. Control*, 89, 2201–2226. DOI: 10.1080/00207179.2016.1151940

Ljung, L. (1999). *System Identification: Theory for the User.* Prentice-Hall, Inc., Englewood Cliffs, NJ, 2nd ed. ISBN: 978-0136566953

Ljung, L. (2022). *System Identification Toolbox® User's Guide.* The MathWorks Inc., Natick, MA. `https://www.mathworks.com/help/pdf_doc/ident/ident_ug.pdf`

Mishra, A. and Bansal K. (2024). Control of two-wheel self-balancing robot: LQR and MPC performance analysis. *2024 IEEE Int. Students' Conf. Electrical, Electronics and Computer Science (SCEECS), 24-25 Feb. 2024, Bhopal, India.* IEEE, 2024. DOI: 10.1109/SCEECS61402.2024.10481856

Simulink Coder (2021). The MathWorks, Inc., Natick, MA. `https://www.mathworks.com/help/pdf_doc/rtw/rtw_ug.pdf`

Petkov, P., Slavov, T., and Kralev, J. (2018). *Design of Embedded Robust Control Systems using MATLAB/Simulink.* IET Control, Robotics and Sensor Series 113. The IET, London. ISBN 978-1-78561-330-2

Raspberry Pi 4 Model B. (2021). Raspberry Pi (Trading) Limited, Cambridge, UK. `https://www.raspberrypi.org/products/raspberry-pi-4-model-b/`

Slavov, T., Petkov, P., Sherif, S., and Kralev, J. (2021). Multi-output identification and LQG control of a two-wheeled robot. *2021 International Conference ICAI, 30 Sept. - 2 Oct. 2021, Varna, Bulgaria.* IEEE. DOI: 10.1109/ICAI52893.2021.9639736

Slavov, Ts., Kralev, J. K., and Petkov, P.H. (2023). Multi-Output Identification and Robust Control of a Two-Wheeled Robot. *9*th International Conference CoDIT, 3-6 July 2023, Rome, Italy. IEEE. DOI: 10.1109/CoDIT58514.2023.10284299

IOP Conf. Series: Materials Science and Engineering **1317** (2024) 012011 doi:10.1088/1757-899X/1317/1/012011

Research on Computer Vision Models for Deep Learning in Autonomous Mobile Robots

Danail Slavov[1*], Anastasiya Slavova[1] and Vladimir Hristov[1]

[1] Department of Electrical Motion Automation Systems, Faculty of Automatics, Technical University of Sofia, Sofia, Bulgaria

*E-mail: d.slavov@tu-sofia.bg

Abstract. We propose a computer vision model for semantic and instance segmentation, based on YOLOv8 architecture and partial convolution modules. Transfer learning on the Cityscapes dataset is performed and results show comparable to the SOTA model accuracy. Furthermore, a speed comparison in terms of inference time is conducted, the results of which show that the proposed model is faster when processing larger input images using smaller batch sizes. Our research enhances the potentialities for further improvement in application of fast deep learning models in mobile robotics for complex vision tasks assisting their intelligent behaviour. It also suggests opportunities to increase the suitability of the improved models when implemented on low-resource hardware, such as edge devices and single-board computers, which in turn can facilitate integration of intelligent features and into various autonomous mobile robots in tasks such as self-driving, avoiding and detouring obstacles, optimal trajectory/path selection, increasing safety and reliability.

1. Introduction

One of the main prerequisites for the implementation of intelligent behavior in autonomous mobile robots (AMR) is the presence of visual functionality, allowing the perception of information from the environment by means of video sensors and its processing by computer systems with the corresponding software. Computer vision is being used with increasing intensity in mobile robotics to perform common or more complex tasks, such as identifying, locating and avoiding/circumventing obstacles to navigate a path to a target, recognizing objects and manipulating them (e.g. picking up, carrying in case of inventory transportation robots) [1] or tracking (e.g. following another robot or a human in the case of collaborative robots). As part of the main aspects of intelligent agents, to which autonomous robots also belong – perception, reasoning, decision making and actuation – vision represents the necessary sense for improving the execution of multifaceted activities without human intervention. Its increased use helps to achieve greater accuracy and speed in various operations, especially with the advance of task specification abstraction toward the goal of increasing robot autonomy in real-world scenarios [2].

In the domain of computer vision, semantic and instance segmentation (IS) approaches outperform the classical object detection or recognition in providing fine-grained details about object boundaries and regions [3] as well as extracting meaningful information about the visual scene. This is especially useful for AMR when performing tasks such as obstacle avoidance, orientation in the environment (for example exact location of pedestrians and other road users in case of self-driving cars), accurately following the road and lanes, passing through openings in production halls, and it also contributes to increase robot's safety and reliability.

Unfortunately, IS is quite a resource-intensive task, requiring an increased number of computational operations, so not all hardware devices typically used in robotics, and especially in mobile, can handle it. In order to overcome this shortcoming, it is necessary to implement lightweight models, which, however, should have a sufficiently good accuracy, allowing reliable performance of the tasks. In this study, the YOLOv8 model is chosen as a reference computer vision system because it is relatively lighter and faster compared to other similar models and also has adequate accuracy. [4] The goal of our research team is to achieve even greater execution speed in inference mode, and subsequently, as part of future development, a smaller memory footprint. This would extend the possibilities of implementing IS models on more restricted hardware that is also cheaper, including embedding applications into various control modules (e.g. for mobile and stationary robots) by integrating these models into the main control system.

2. Related work

YOLOv8 is considered a state-of-the-art (SOTA) deep learning model for real-time classification, object detection and image segmentation in computer vision applications. Its efficient architecture and algorithms as well as ability to identify and localize objects in real-time scenarios enable a wide usage in industries such as robotics and autonomous driving. Starting from YOLOv1, subsequent versions further refined the model's capabilities. These iterations introduced improvements in terms of accuracy and speed, incorporating techniques such as anchor boxes, feature pyramid networks, and multi-scale prediction for enhanced performance.[5] The latest version offers pre-trained models on the COCO (Common Objects in Context) dataset[6] with 80 classes and options for building custom models for a specific computer vision task. It introduces the C2f block (Fig. 1) with optimized convolutions, anchor-free detection that predicts bounding boxes at the center of objects automatically and stopping mosaic augmentation for overfitting prevention. The structure of YOLOv8 for instance segmentation is presented in Fig. 2. [7]

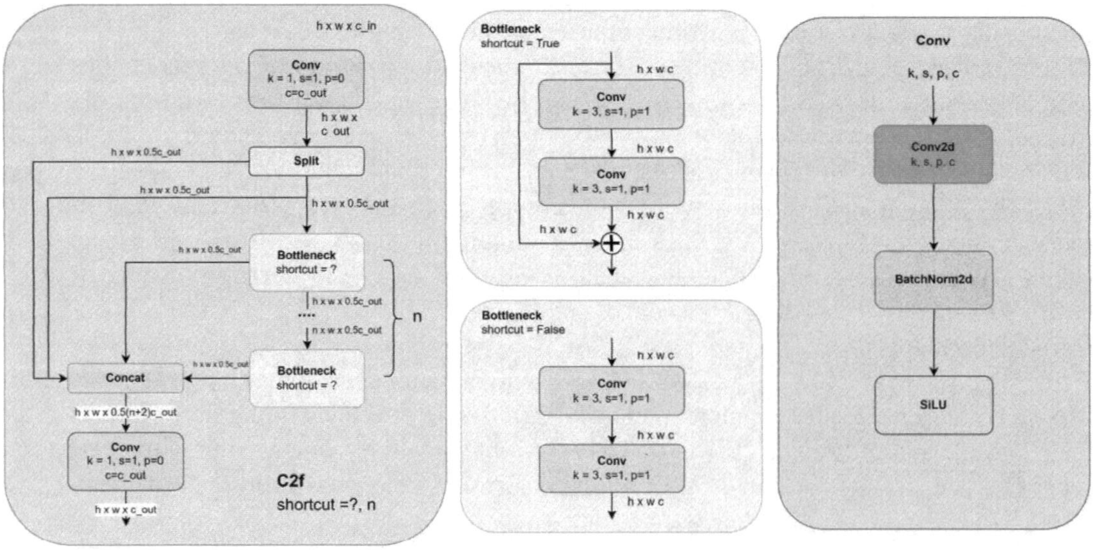

Figure 1. Building blocks of YOLOv8.

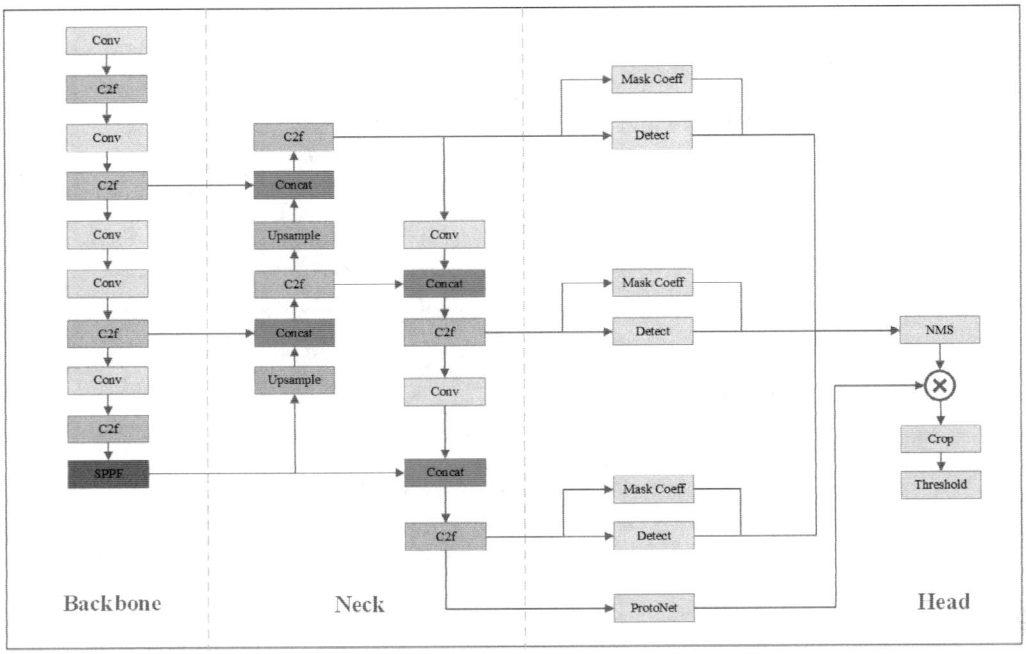

Figure 2. YOLOv8 segmentation structure.

It uses a segmentation head as in [8] and [9] which adds a mask branch to an existing one-stage object detection model in the same way as Mask R-CNN [10] does to Faster R-CNN [11] avoiding the feature localization step. The complex task of instance segmentation is divided into two simpler, parallel tasks that can be assembled to form the final masks. This way, prototypes and mask coefficients are computed independently and then assembled as a single matrix multiplication, which maintains spatial coherence and speed.

Because of the task's high computational requirements, most real-time models for semantic and instance segmentation are to some extend unsuitable for low-resource embedded devices. Much work has been done to design efficient and lightweight models capable of running on resource-constrained devices, such as NVIDIA Jetson and even Raspberry Pi. One of them is the JetSeg project [12] which, among other improvements, uses a novel operator for lightweight and efficient feature extraction, called JetConv (Fig. 3, left). It introduces a so called Combinated Dilated Depthwise Convolution (CDDC) which utilizes dilated asymmetric convolution followed by a non-asymmetric and non-dilated convolution. This enables effective and balanced spatial information representation and increases the receptive field, which allows capturing long-range dependencies without increasing the total number of parameters and balancing between power and computational efficiency.

Another simple yet fast and effective block described in [9] aims to increase the value of FLOPS by lowering the frequency of convolution operations memory access (Fig. 3, right). For this purpose, the authors create an operator performing a partial convolution on a fraction of the feature maps. The efficiency is optimized by leveraging the feature maps' redundancy as they share high similarities among different channels [13] [14]. Proposed PConv module aims to simultaneously reduce computational redundancy and memory access. It applies a regular convolution on only a part of the input channels for spatial feature extraction and leaves the remaining channels untouched. The first or the last consecutive portion of channels are considered representative of the whole feature maps for computation. The input and output feature maps' shape are the same in terms of both spatial dimensions and number of channels. In this way, with a partial ratio of 1/4, PConv has only 1/16 FLOPs and 1/4 memory access of a regular convolution.

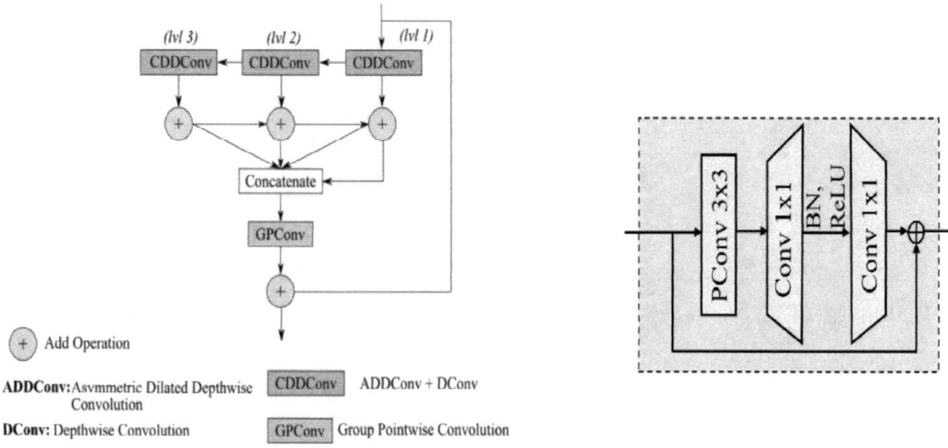

Figure 3. JetConv (left) and FasterNet (right) blocks.

3. Proposed method and structure

For our modified neural network we use the implementation of YOLOv8, developed by Ultralytics. [15] It is written in the Python programming language using the machine learning library PyTorch. The AI framework supports multiple computer vision tasks, such as detection, segmentation, classification, and pose estimation. Furthermore, the model has nano, small, medium, large and extra-large variants with different depth, width and ratio parameters (Table 1).

Table 1. Multiplication parameters of YOLOv8 variants.

Model variant	Depth multiple	Width multiple	Max. channels
n	0.33	0.25	1024
s	0.33	0.50	1024
m	0.67	0.75	768
l	1.00	1.00	512
x	1.00	1.25	512

The depth multiple determines the number of bottleneck blocks in the C2f block, while the width multiple and the ratio determine the number of output channels of each layer. Due to decoupled head approach the model structure can adapt to each of the supported tasks and in our case this is instance segmentation.

In an effort to further optimize model's efficiency and increase both training and inference speeds, instead of the original „Conv" block we incorporate a new convolution block into the YOLOv8 backbone and neck. It is inspired from the above described FasterNet model with several changes. In the original YOLOv8 structure, each „Conv" layer reduces the spatial size by half and either doubles the output number of channels (in the backbone) or leaves it unchanged (in the neck). On the other hand, the original FasterNet block retains all three dimensions (width × height × channels) unchanged. To make it capable of being integrated into the YOLO architecture, after summation in FasterNet block we add an option for another point convolution which can expand the number of output channels, followed by a max-pooling layer with kernel size 2 × 2 for spatial reduction. Then we replace the P3, P4 and P5 Conv in the backbone as well as both Conv blocks in the neck with the new FNBlock, while leaving P1 and P2 unchanged. Table 2 shows the consecutive layers and blocks with their parameters in the modified neural network structure. The first two modules with full convolutions from the

IOP Conf. Series: Materials Science and Engineering **1317** (2024) 012011 doi:10.1088/1757-899X/1317/1/012011

original structure are assumed to preserve its high feature extraction ability while lightweighting some of the next layers aims to improve model's speed. Our rationale is based on

Table 2. List structure of the modified model (index -1 indicates previous block; SPPF – Spatial Pyramid Pooling Fast).

Index	From	Repeats	Block Name	Arguments		
	Backbone					
0	-1	1	Conv	64	3	2
1	-1	1	Conv	128	3	2
2	-1	3	C2f	128	shortcut = True	
3	-1	1	FNBlock	256	3	2
4	-1	6	C2f	256	shortcut = True	
5	-1	1	FNBlock	512	3	2
6	-1	6	C2f	512	shortcut = True	
7	-1	1	FNBlock	1024	3	2
8	-1	3	C2f	1024	shortcut = True	
9	-1	1	SPPF	1024		
	Neck					
10	-1	1	Upsample	None	2	nearest
11	-1, 6	1	Concat			
12	-1	3	C2f	512	shortcut = False	
13	-1	1	Upsample	None	2	nearest
14	-1, 4	1	Concat			
15	-1	3	C2f	256	shortcut = False	
16	-1	1	FNBlock	256	3	2
17	-1, 12	1	Concat			
18	-1	3	C2f	512	shortcut = False	
19	-1	1	FNBlock	512	3	2
20	-1, 9	1	Concat			
21	-1	3	C2f	1024	shortcut = False	
	Head					
22	15, 18, 21	1	Segment	output channels = number of classes		

the documented advantages of the partial convolution: while utilizing commonly present redundancy in feature maps among the subsequent layers of a convolution neural network, it also lowers both computational burden and frequent memory access towards the hardware device.

Because the current research is focused on mobile robots, including AMR used in industry settings, unmanned vehicles and self-driving cars, we decided to test the IS capabilities of the modified model in urban street scenes involving multiple road users, such as cars, trucks, buses, cyclists, motorcyclists and pedestrians. A representative sample of such a subject area is implemented in the Cityscapes dataset [16], specifically designed for training, validating and testing models for semantic and instance segmentation. This large-scale dataset contains a diverse set of stereo video sequences recorded in street scenes from 50 different cities, with high quality pixel-level annotations for 30 classes of objects. The part we use leftImg8bit consists of 3457 images from 18 cities in Germany with 2965 (86%) training and 492 (14%) validation color images. Annotations are in the form of color images and represent masks where the pixels of each segment are in the color corresponding to their class. These .png files are

IOP Conf. Series: Materials Science and Engineering **1317** (2024) 012011 doi:10.1088/1757-899X/1317/1/012011

converted to .json files for Mircosoft's COCO Format and then to .txt files for the YOLO segmentation format [17].

4. Experiments and results

Transfer learning on the Cityscapes datasets is conducted using model states pre-trained on COCO dataset which are provided by Ultralytics. First, the original unchanged YOLOv8 segmentation model is trained in its "nano" and "small" variants (yolov8n-seg and yolov8s-seg respectively). We settled on the two smallest variants because of our goal to explore the performance capabilities on low-resource hardware. Each of them is trained with two sizes of the input images: 640 × 320 (close to the default size for which the pre-trained models are optimized) and 2048 × 1024 (the size of Cityscapes images). Modified structures are created in the manner described in Chapter 3, and the parameters of the unmodified layers are inherited from the previously trained on COCO models. Transfer learning is again performed with both input sizes. Results for the "small" architectures showing segmentation loss improvement measured over the training and the validation sets are presented in Fig. 4. We also explore the increase of the mAP(50-95) metrics [18] [19] with the advancement of the training epochs, as seen on Fig. 5.

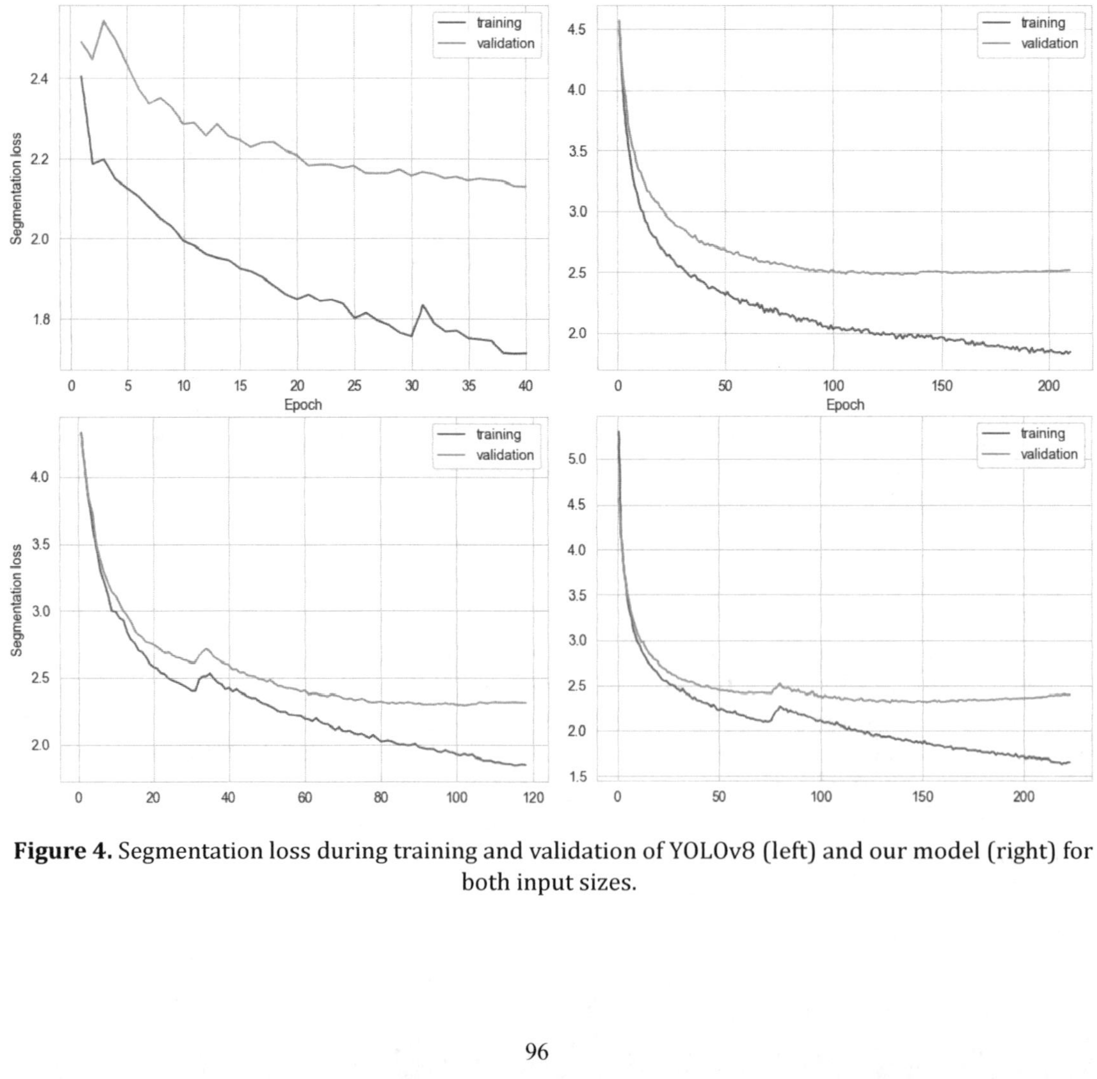

Figure 4. Segmentation loss during training and validation of YOLOv8 (left) and our model (right) for both input sizes.

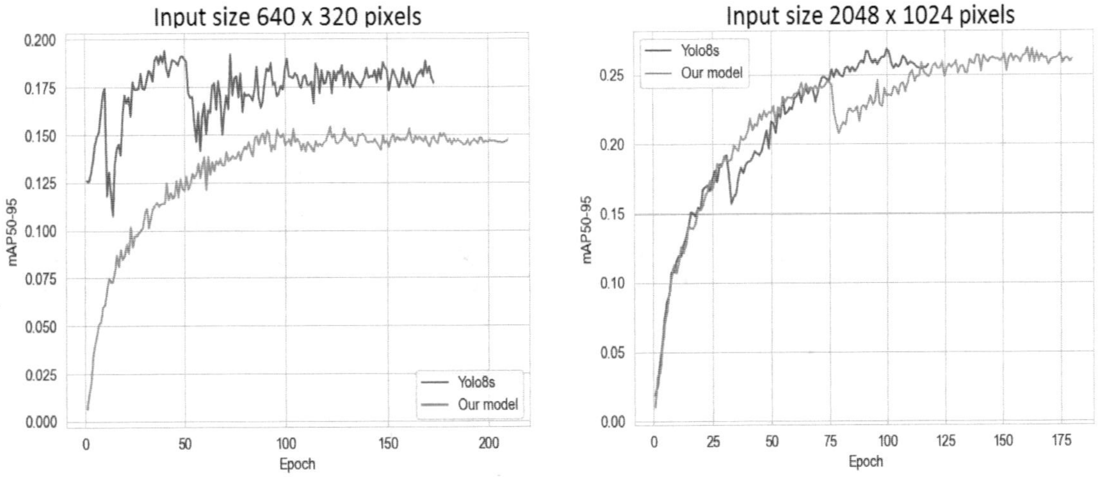

Figure 5. Training of "small" variants for two input sizes (mAP(50-95) per epoch).

As the graphics show, both original and modified models achieve lower metrics with the smaller input data size. The reason for this could be sought in the image resizing process which weakens the presence of smaller details in the urban scenes and subsequently, their contribution toward model generalization abilities decreases. When both models are trained with the original image size, they achieve similar accuracy around 0.27 in mAP(50-95). The original variant reaches this value faster because it is pre-trained to a higher degree – all its trainable parameters are inherited from a saved state. While in the modified models there is a total of 45 parameter tensors which need to be trained from scratch. The "nano" variants achieved slightly poorer metrics altogether which suggests that input sample fitting into a smaller network, especially in its depth where semantics are stored, is more difficult.

Upon training completion, a comparison of the speed in inference mode is made between the different variants at several batch sizes with three sizes of input images: 320 × 160, 640 × 320, and 2048 × 1024. The performance is explored on GPU (NVIDIA GeForce RTX 4090, 24 GB). The results for the speed measurements are presented in Figures 6 and 7.

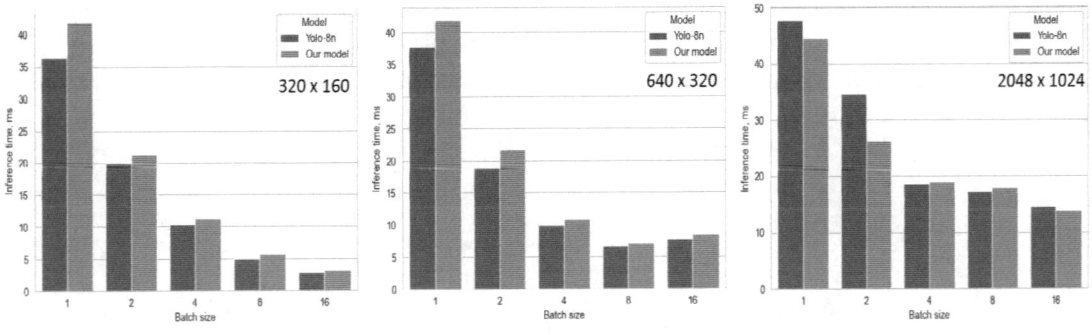

Figure 6. Inference time for the "nano" structures.

Although time values for both unmodified and modified models are close at most of the batch sizes, a trend is emerging that the smaller the input data size, the faster the original

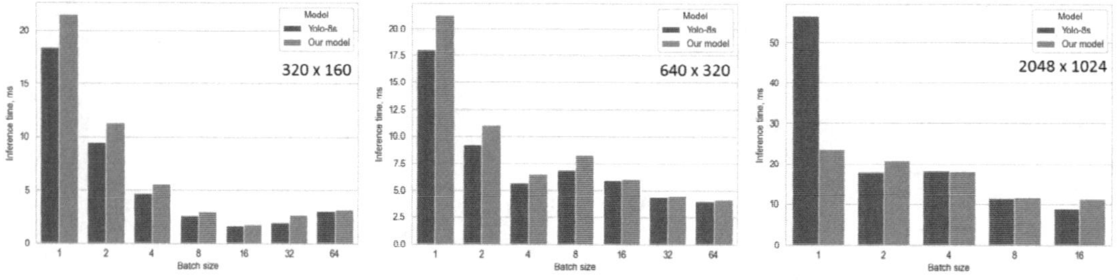

Figure 7. Inference time for the "small" structures.

YOLOv8 models are, as their structure and efficiency are optimized mainly for processing smaller images. With the "nano" structure, our model is faster with the largest images in almost all batch sizes but with "small", it is faster – although significantly – only at batch size 1. Results suggest that the modified variants have improvement potential to reach even faster inference speeds, especially in devices with low-resource GPUs where smaller batch sizes are preferable or the only reachable.

Measurements of additional metrics about model efficiency and its time and space complexity are registered, including the total floating point operations (FLOPs) and number of parameters contained in the entire network (Table 3).

Table 3. Space and time performance metrics.

Model		Layers	Parameters	GFLOPs
YOLOv8-seg	nano	261	3267126	12.1
	small		11797062	42.7
Our model	nano	287	2867830	12.3
	small		10197830	43.3

As seen from the table, our modified networks have greater number of layers and slightly higher FLOP count. This indicates a certain redundancy, introduced by our modifications, namely the additional point convolution layer in the FNBlock which aims to control the output channel number. On the other side, the number of trainable parameters is smaller which can be attributed to the reduced number of filters in the partial convolution compared to the standard one.

5. Conclusion

In this work, we research semantic and instance segmentation models suitable for computer vision tasks in AMRs. We incorporate partial convolution modules into YOLOv8 structure and explore the resulting benefits. Transfer learning on large images from urban scenes shows almost no loss of accuracy by the proposed models. Furthermore, their inference speeds are higher under certain conditions, especially when processing larger input sizes and lower batch sizes. This makes them more appropriate for cases where high resolution of the visual perceptions are required and also for devices with lower computational capabilities which are typically embedded into mobile robots. The results obtained allow for further refinement and adaptation to tasks performed by AMR including self-driving vehicles which intelligent behaviour is assisted even by the smaller visual details. Of interest for future research would be the additional optimization of the structure, and hence the performance, to adapt the model to even more resource-limited devices, such as Raspberry Pi, which do not have specialized graphic processors. The further research is planned to include exploration of more variants of the modified architecture in order to achieve faster inference speeds, especially on CPU.

Acknowledgements

The authors would like to thank the Research and Development Sector at the Technical University – Sofia for the financial support in this conference and contract № 242ПД0019-08 as well as contract № 241ПР0014-08.

References

[1] https://hy-tek.com/resources/types-of-autonomous-robots, 2024-05-06
[2] Isabel M. Rayas Fernández, 2023, arXiv:2307.13163
[3] https://www.picsellia.com/post/segmentation-vs-detection-vs-classification-in-computer-vision-a-comparative-analysis, 2024-05-07
[4] https://roboflow.com/model-task-type/instance-segmentation, 2024-03-10
[5] https://keylabs.ai/blog/under-the-hood-yolov8-architecture-explained/, 2024-02-01
[6] https://cocodataset.org/#home, 2024-05-10
[7] https://github.com/Johnny-zbb, 2024-05-22
[8] Daniel Bolya, Chong Zhou, Fanyi Xiao, Yong Jae Lee, 2019, arXiv:1904.02689v2
[9] Jierun Chen, Shiu-hong Kao, Hao He, Weipeng Zhuo, Song Wen, Chul-Ho Lee, S.-H. Gary Chan, 2023, arXiv:2303.03667v3
[10] Kaiming He, Georgia Gkioxari, Piotr Doll´ar, Ross Gir-shick, 2017, IEEE International Conference on Computer Vision (ICCV), 2980-2988, 322
[11] Shaoqing Ren, Kaiming He, Ross Girshick, Jian Sun, 2015, NeurIPS
[12] Miguel Lopez-Montiel, Daniel Alejandro Lopez, Oscal Montiel, 2023, arXiv:2305.11419v1
[13] Kai Han, Yunhe Wang, Qi Tian, Jianyuan Guo, Chunjing Xu, Chang Xu, 2020, IEEE/CVF conference on computer vision and pattern recognition, pages 1580–1589
[14] Qiulin Zhang, Zhuqing Jiang, Qishuo Lu, Jia'nan Han, Zhengxin Zeng, Shang-Hua Gao, Aidong Men, 2020, arXiv:2006.12085
[15] https://github.com/ultralytics/ultralytics, 2024-03-15
[16] https://www.cityscapes-dataset.com/, 2024-05-10
[17] https://docs.ultralytics.com/datasets/segment/, 2024-05-10
[18] https://www.v7labs.com/blog/mean-average-precision, 2024-05-15
[19] https://jonathan-hui.medium.com/map-mean-average-precision-for-object-detection-45c121a31173, 2024-05-15

AUTHOR INDEX